INSTRUCTOR'S MANUAL

to

EUCLIDEAN GEOMETRY:
a first course

INSTRUCTOR'S MANUAL

to

EUCLIDEAN GEOMETRY:
a first course

by Mark Solomonovich,

Grant MacEwan University

Instructor's Manual to Euclidean Geometry

A First Course

iUniverse books may be ordered through booksellers or by contacting:

iUniverse
1663 Liberty Drive
Bloomington, IN 47403
www.iuniverse.com
1-800-Authors (1-800-288-4677)

ISBN: 978-1-4502-5785-5 (sc)
ISBN: 978-1-4502-5786-2 (ebook)

Printed in the United States of America

iUniverse rev. date: 9/24/2010

CONTENTS

ADDITIONAL SOLUTIONS

for

EUCLIDEAN GEOMETRY: a first course by M. Solomonovich.

Section 1.2

13. Newton's laws of motion, which are the axioms for Galilean-Newtonian Mechanics, contain the notions of *force* and *mass*. One can assume either of them (but only one of them) to be an undefined notion of Mechanics; then the other one can be defined from the second law of motion, which can be formulated accordingly in two ways:

(1) The force applied to a body (material point) is the product of its mass and acceleration. (This formulation defines the notion of *force* in suggestion that *mass* is an undefined notion).

(2) The acceleration experienced by a body (material point) is proportional to the force applied to the body; the coefficient of proportionality is called the mass of the body . (Here the notion of *force* is undefined).

It is more convenient to assume *force* to be the undefined notion, since otherwise it is getting tricky to formulate the first law of motion, which also includes the notion of *force*:

If no force acts on a body, it will remain at rest or keep moving with constant velocity.

Section 1.3

15. First of all, we assume that lines that are labeled by different letters are distinct lines.

Two distinct lines can have at most one common point, as follows from Axiom 1; hence lines *a* and *b* have only one common point; let us name it *P*. Line *c* also passes through P, since it is concurrent with *a* and *b*.

Now let us show that line *d* must necessarily pass through *P*. If *d* does not pass through *P*, it must have another common point with lines *a* and *b*, which is impossible since these two lines have only one common point *P*. Hence, *d* also passes through *P*, and therefore all four lines are concurrent.

16. It is given that the line passing through *A* and *B* (by Axiom1 there is only one such line), contains *O*. The line through *A* and *C* contains *O* as well. Since there is only one line through *A* and *O*, this is the line into which *AB* and *AC* are being extended; thus *A*, *B*, and *C* are collinear.

22. The diagrams below show possible respective positions of the points: *M* can lie in the interior (diagram (a)) or exterior (diagram (b)) of *RS*. These situations allow five possible positions of *N* (shown by arrows), since *R* cannot be located on *NS*.

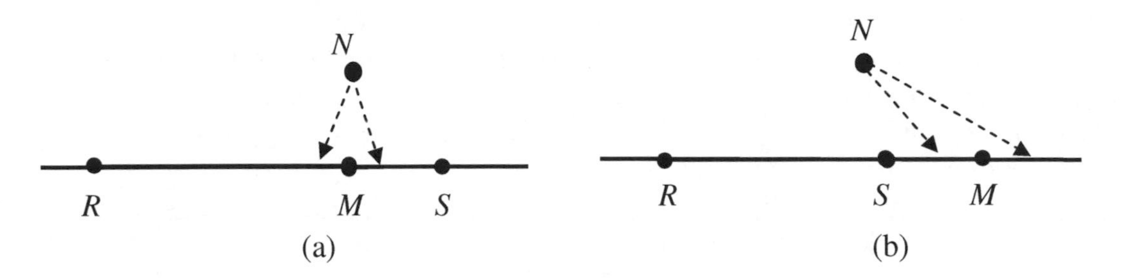

Each configuration of the points can be seen as a *word* in which the letters follow from the left to the right (from West to East), and we shall call two configurations different if they correspond to different *words*. Then there are 4 different configurations shown in the above diagrams (one for each possible location of N). For each of these configurations there is another one, represented by the same word written from the right to left (East to West); so there are 8 different configurations.

If it is not required that $MN < MS$, point N can be located to the East from S on diagram (a) and between S and R (closer to S) on diagram (b), and then four more configurations will be possible.

30. We suggest, without losing generality, that $b \geq a$.

(a) The midpoints of OA and OB are located at the distances $OP = \frac{1}{2}a$ and $OQ = \frac{1}{2}b$ respectively from O. Then the distance between the midpoints is $PQ = \frac{1}{2}b - \frac{1}{2}a = \frac{b-a}{2}$.

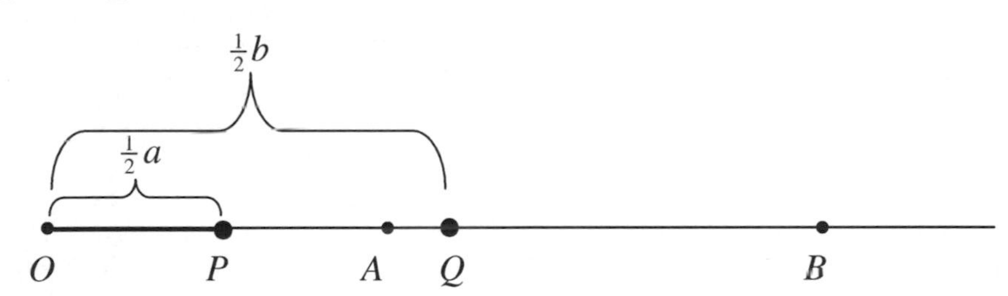

(b) $OM = OA + AM = OA + \frac{1}{2}(OB - OA) = \frac{1}{2}a + \frac{1}{2}(b-a) = \frac{1}{2}(b+a)$.

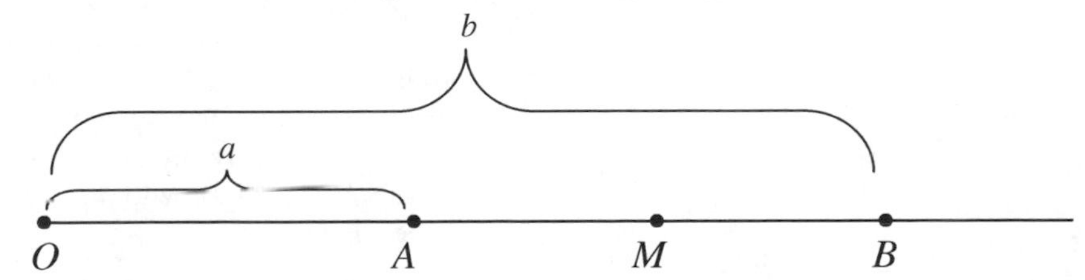

37. A few examples of transformations applied to a plane follow below. The transformations and the corresponding results of applying these transformations to the plane are labeled by Roman letters I – VIII. The plane is shown on the diagrams as a square, though we understand that a plane extends indefinitely in all directions (one may say: it is infinitely long and infinitely wide). A brief description of every transformation applied to the original plane (which is denoted **I**)is given below the corresponding image of the plane.

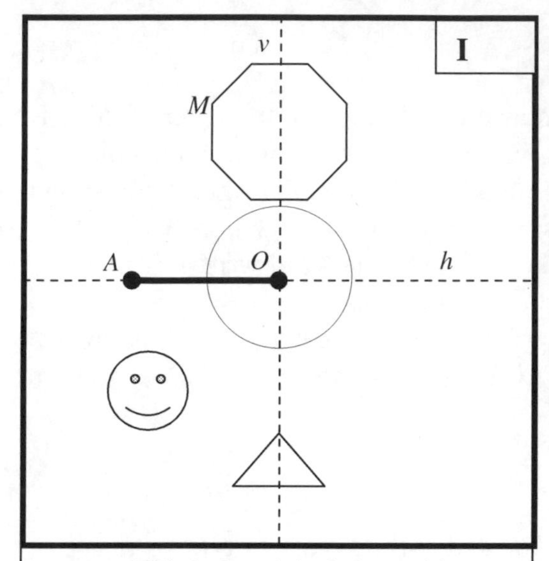

This is the original plane with a few figures (circle, hexagon, triangle, face) and two lines: vertical (v) and horizontal (h).

The lines intersect at point O, the centre of the circle. We suggest that the positions of the hexagon and the triangle are *symmetric with respect to line v* (what that means will be explained in the comment to transformation **II**).

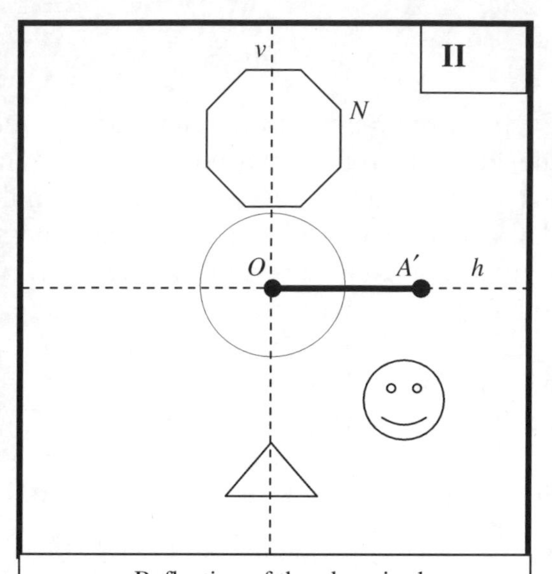

Reflection of the plane in the vertical line v. This is a type (iii) isometry that leaves every point of line v unchanged. It transforms line h onto itself (we can say: *h is invariant* with respect to this transformation, or *h is mapped onto h*), however every point of line h except point O changes its position . For example, A is mapped into A', and segment OA' is the image of OA under this transformation.

Due to the property that we shall call *axial symmetry* of the circle, hexagon, and triangle, and their symmetric positions with respect to the line v, they do not change (*are invariant*) under the transformation, although particular points of these figures may change their position: for example the vertex M of the hexagon is mapped into N.

The face is not invariant under this transformation, since it was not positioned symmetrically with respect to v.

One may say that the face has changed its position with respect to the other figures.

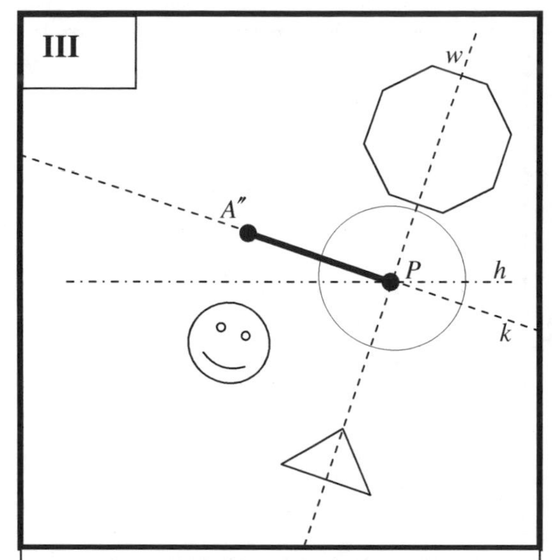

This transformation is a type (i) isometry that translates point O into a new position, at some point P located on the line h. Thus P is the *image* of O under this transformation.

Since the transformation is an isometry, none of the other points of the plane can keep its previous position (explain in detail, why). At the same time, the sizes of the figures, such as the radius of the circle or the side of the hexagon, and the distances between them will not change.

We shall assume without a proof that an isometry transforms a straight line into a straight line. (This will be proved later, based on the results of section 4.7). So both h and v will be transformed into straight lines (labeled k and w respectively in the diagram).

We know that point P is located on h; however it is not guaranteed by our axioms that line h will be transformed into itself: its *direction* (whatever this word means) may change. We shall be able to explain the word *direction* and introduce the so-called *parallel translations* in Chapter 6; so far our translations are not parallel: they can only deliver a given point into a required point.

Yet, if we wish to perform an isometry that transforms h into itself, we can achieve this by rotating the plane about P. This is done by the next transformation (see **IV**).

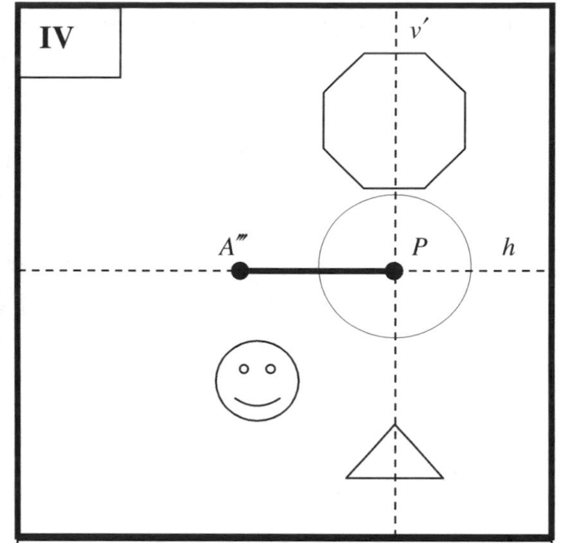

This transformation is such a rotation of the plane about point P that line k will take the position occupied by the line h before the previous transformation (**III**). This is a type (ii) isometry.

This new plane (**IV**) is obtained as a result of applying two consecutive isometries to the original plane (**I**): a translation that carried O into P has been followed by the rotation about P that made line k coincide with h. We say in cases like this that a *composition* of two transformations has been applied to the plane.

The respective positions of the figures in (**IV**) are the same as in the original plane (**I**).

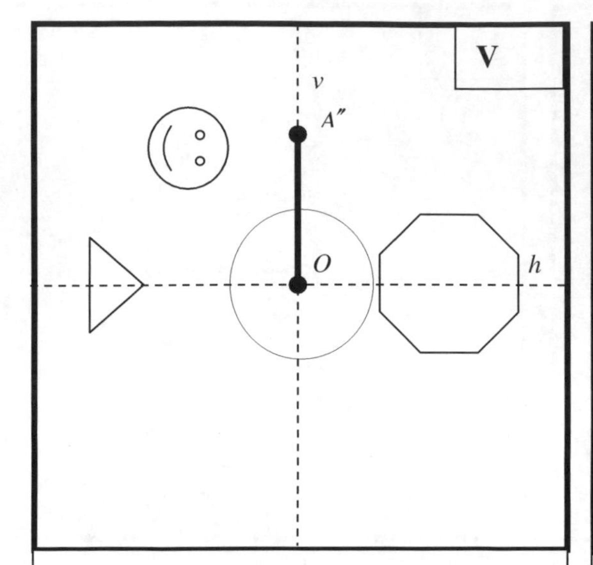

Describe how this isometry can be performed. Which types of motions are to be used? (Propose a few solutions).

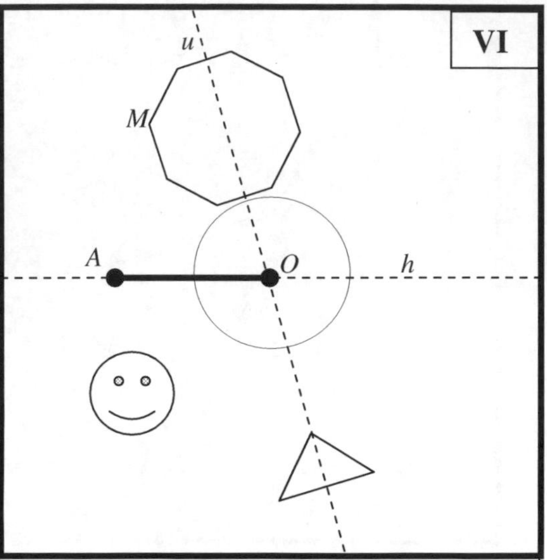

This transformation is not an isometry, even though the figures have preserved their shapes. Explain why it is not an isometry. (Hint: consider the distance *MA*)

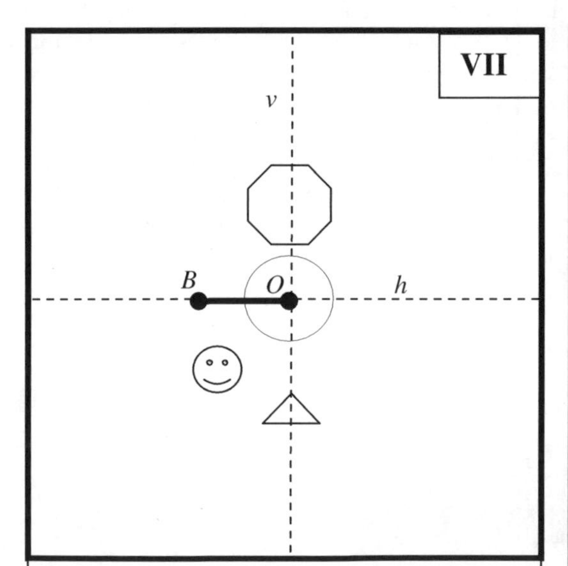

The plane has been contracted evenly in all the directions towards point *O* (in Chapter 9 we shall call such a transformation a homothety). This is not an isometry since every segment in the plane has become less than it was.

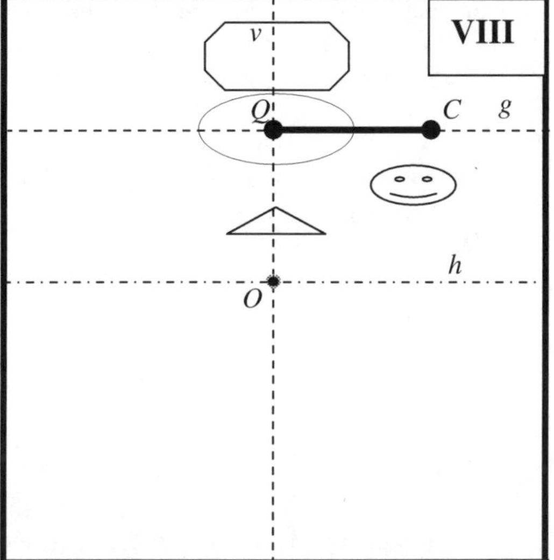

This is a composition of a contraction in the vertical direction only, translation, reflection, and, probably, a rotation. It is not an isometry.

Is it always so that a composition of non-isometries is not an isometry? Can a composition of non-isometries be an isometry? (Hint: Yes, it can).

44. Let us introduce the following notations for the various types of transformations we may need:
T – translation that carries a point into a point (type (i) isometry);
ρ – rotation about a point (type(ii) isometry);
R – reflection in a line (type (iii) isometry);
S – stretching (stretching is not an isometry as well as its inverse, called contraction, since these transformations change the sizes of figures. Such a stretching or a contraction that does not change the shape of figures but only their sizes, is called a *homothety*; these will be considered in Chapter 9, when studying the *similarity* of figures).

a) A translation along a line (one step forward), followed by a rotation in case the feet deviates from the original direction.
Formulae: T(big left) or $\rho \circ T$ (big left).
b) A translation (maybe combined with a rotation) and a reflection; the reflection may follow or precede the translation.
Formulae: $R \circ T$ (big left) or $R \circ \rho \circ T$ (big left); also $T \circ R$ (big left) or $\rho \circ T \circ R$ (big left), etc.
c) This must be a combination of at least the following three transformations: translation along a line, reflection, and contraction; the order of these may be any. Rotation may be added at some step if, for instance, the footprint deviates from the original direction after the translation.
Formulae: $S^{-1} \circ R \circ T$ (big left) or $R \circ T \circ S^{-1}$ (big left) or ...
d) The inverse of the transformations listed in (a) are T^{-1}(forward big left) and $(T \circ \rho)^{-1} = T^{-1} \circ \rho^{-1}$ applied to the forward big left footprint.

Let us notice that the inverse of a product (composition) of transformations is equal to the product of the inverses in the opposite order: the transformation that has been performed the last will be the first to invert, etc.

For the first two transformations in (b) the inverses are $(T \circ R)^{-1} = R^{-1} \circ T^{-1}$ and $(R \circ \rho \circ T)^{-1} = T^{-1} \circ \rho^{-1} \circ R^{-1}$, both applied to the big right footprint.

For the first transformation in (c), the inverse is $\left(S^{-1} \circ R \circ T\right)^{-1} = T^{-1} \circ R^{-1} \circ S$; it should be applied to the small right footprint.

45. Let us show that a composition of isometries is an isometry (the word *show* in mathematics is a synonym of the word *prove*).

Let transformation τ be a composition of transformations α and β: $\tau = \beta \circ \alpha$.

It means that in order to apply τ to a plane, the plane should be subjected to transformation α followed by β. If A is a point in the plane, then

$$\tau(A) = \beta \circ \alpha(A) = \beta(\alpha(A)),$$

which means that in order to obtain the image of A under transformation $\tau = \beta \circ \alpha$, one should apply transformation β to the image of A obtained as a result of transformation α.

Let A and B be some points in a plane, which is subjected to α followed by β. We denote their images after the first transformation A' and B', and the images of A' and B' after the second transformation has been applied will be denoted A'' and B'':

$$\begin{aligned} &A' = \alpha(A); \quad B' = \alpha(B) \\ &A'' = \beta(A'); \quad B'' = \beta(B') \end{aligned}.$$

Then A'' and B'' can be viewed as the respective images of A and B as a result of the composition of α and β

$$\left.\begin{aligned} A'' = \beta(A') = \beta(\alpha(A)) = \tau(A) \\ B'' = \beta(B') = \beta(\alpha(B)) = \tau(B) \end{aligned}\right\}.$$

Since α is an isometry, $AB = A'B'$, and since β is an isometry, $A'B' = A''B''$; therefore $A''B'' = AB$,

or $\tau(A)\tau(B) = AB$: as a result of transformation $\tau = \beta \circ \alpha$, segment AB is transformed into a segment congruent to AB. We did not make any special assumptions about the points A and B, they are arbitrary points; therefore we can say that $\tau = \beta \circ \alpha$ transforms every segment of the plane into a congruent segment. According to the definition of isometries, the latter means that τ is an isometry.

Similarly, by considering the segments formed by images and preimages of points in a plane subjected to an isometry, one can show that the transformation inverse to an isometry is an isometry as well.

54. It is convenient to base the solution on the following theorem:

Theorem. $m \equiv n(\bmod k)$ **if and only if the difference** $(m-n)$ **is divisible by** k.

Proof. By definition, $m \equiv n(\bmod k)$ if the numbers m and n produce the same remainder r when divided by k, i.e. $m = kp + r; \quad n = kq + r$, where p and q are some integers and r is some nonnegative (positive or 0) integer that is less than k: $0 \le r < k$. Then $m - n = kp + r - (kq + r) = k(p-q)$, which means that the difference between m and n is divisible by k.

Now let us suggest that $(m-n)$ is divisible by k, i.e. there exists such an integer j that $m - n = kj$. Let R and r denote the remainders of m and n respectively when these numbers are divided by k: $m = kp + R; \quad n = kq + r$, where p and q are some integers. Each of these remainders is a nonnegative number that is less than k: $0 \le R < k; 0 \le r < k$.

Then $m - n = k(p-q) + (R-r)$. Since this difference is divisible by k, and the addend $k(p-q)$ is divisible by k as well, the difference between the remainders $(R-r)$ must be divisible by k. This is impossible if the difference is not 0, since each of the

numbers R and r is nonnegative and less than k, which means that the absolute value of their difference is less than k.
Therefore, the remainders are equal, Q.E.D.

a) Now let us apply the result of this theorem in order to prove that *congruence modulo k* is an equivalence relation on the set of integers.
(i) $m-m=0$, which is divisible by any number, including k; thus $m \equiv m(\text{mod}\, k)$, i.e. Reflexivity holds.
(ii) If $m \equiv n(\text{mod}\, k)$, i.e. $(m-n)$ is divisible by k: $m-n=kp$, where k is some integer, then $n-m=k(-p)$, which means that $(n-m)$ is also divisible by k; hence $n \equiv m(\text{mod}\, k)$, and Symmetry holds.
(iii) If $m \equiv n \,\text{mod}(k)$ and $n \equiv p \,\text{mod}(k)$, then $\left.\begin{matrix} m-n=ki \\ n-p=kj \end{matrix}\right\} \Rightarrow m-p=k(i+j)$, which means the divisibility of $(m-p)$ by k and therefore the congruence of m and p modulo k. Therefore, Transitivity holds.
The three properties of equivalence relations hold, therefore the *congruence modulo k* is an equivalence relation, Q.E.D.

b) Two integers that produce the same remainder when divided by 5, will be members of the same equivalence class; therefore there will be an equivalence class for each distinct remainder.
The division by 5 can produce only 5 different remainders: 0, 1, 2, 3, and 4; each of them defines an equivalence class:
[0] = {..., -15, -10, -5, 0, 5, 10, ..., $5n$, ... }
[1] = {..., -14, -9, -4, 1, 6, 11, ..., $5n+1$, ...}
[2] = {..., -13, -8, -3, 2, 7, 12, ..., $5n+2$, ...}
[3] = {..., -12, -7, -2, 3, 8, 13, ..., $5n+3$, ...}
[4] = {..., -11, -6, -1, 4, 9, 14, ..., $5n+4$, ...}.

Section 2.1

13. A statement is called a proposition if it can be proved based on results that are believed to be true, such as axioms or other statements following from the axioms..
a) This is a proposition following from the properties of real numbers.
b) This is a proposition following from the properties of real numbers.
c) This is a proposition following from the Archimedean law of buoyancy.
d) This statement can be viewed as a proposition following from the energy conservation law; yet sometimes this statement is deemed as another formulation of the law of conservation of energy; in the latter case we can call it an axiom, not a proposition.
e) It is a particular case of the definition of a natural power.
f) This is a definition.

Section 2.2-3

3.

a) (iii) is the negation.

(i) denies the statement (if it is a cat then it cannot be a dog), but its negation will not be the original statement (if it's not a cat it may be still some other than dog animal, e.g. a horse). Thus (i) does not complement the given statement to tautology (universally true statement). (One may also say that if (i) were the negation, the law of the excluded middle would be violated).

(ii) is a negation of some other statement, not of the given one.

b) (ii) is the negation.
(i) does deny the statement but it does not include all possible situations in which the statement is denied, in particular the expression may be equal to zero as`well.
(iii) does not deny the statement: the expression in brackets may be positive even if the expression outside them is positive.

c) (iii) is the negation.
(i) does deny the statement , but it does not complement the statement to universal truth: there are numbers that are not divisible by 5 and not prime at the same time, e.g. 6, 8, 9, etc.
(ii) does not complement the statement to universal truth: it does not include prime numbers for example (one may say that it does not include any numbers since a number that is divisible by all factors but 5 does not exist: the number of prime factors is infinite).

d) (iii) is the negation.
(i) does not complement the statement to universal truth.
(ii) is completely irrelevant to the statement: the congruence of segments does not depend on whether the segments do lie in one line or they do not.

Sections 4.6-7

15. Let *AM* be a median drawn from *A* to the midpoint *M* of *BC* in some triangle *ABC*. It follows from the triangle inequality that in ΔAMB, $AM < AB + \frac{1}{2}BC$; similarly, in ΔAMC, $AM < AC + \frac{1}{2}BC$. After adding these two inequalities, we obtain: $2AM < AB + BC + AC$, which means that a median is less than half-perimeter, Q.E.D.

4.REVIEW

2. Let ΔABC be isosceles, with *AB=BC*, and *AD* and *CE* be the altitudes dropped onto the congruent sides, as shown in diagram (a), or dropped on their extensions if the angle at the vertex (also called the vertical angle) is obtuse as in diagram (b) .

If the vertical angle is right, the altitudes coincide with the legs, which makes the conclusion true since the lateral sides (legs) are congruent by the hypothesis.

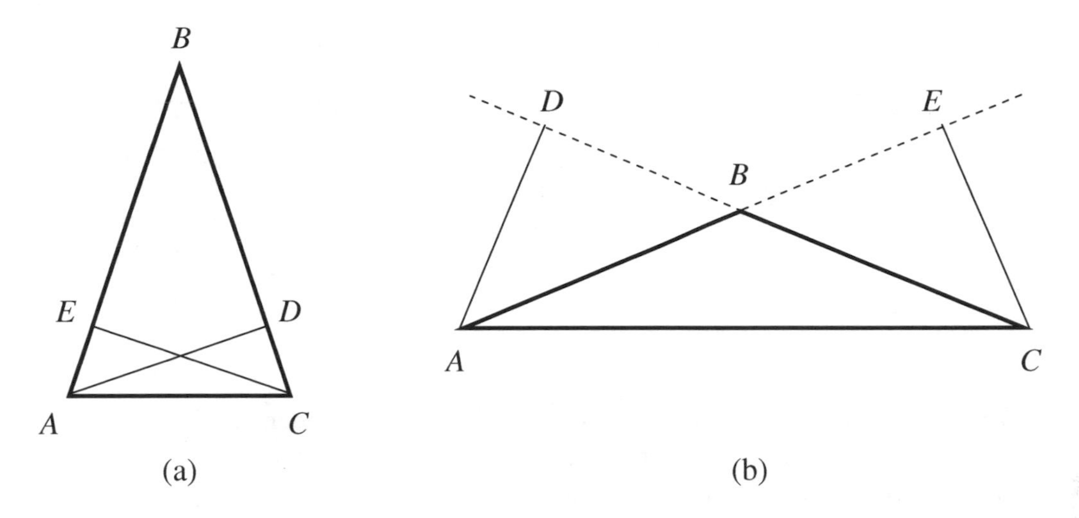

(a) (b)

$\angle BAC = \angle BCA$ as the base angles in an isosceles triangle, and AC is the common hypotenuse in right triangles ADC and CEA; therefore these triangles are congruent by the hypotenuse and an acute angle. Hence, $AD=CE$ as corresponding legs in congruent right triangles, Q.E.D.

3. Let ΔABC be isosceles, with $AB=BC$, and AM and CN be the bisectors drawn from the vertices opposed the congruent sides, as shown in the diagram below.

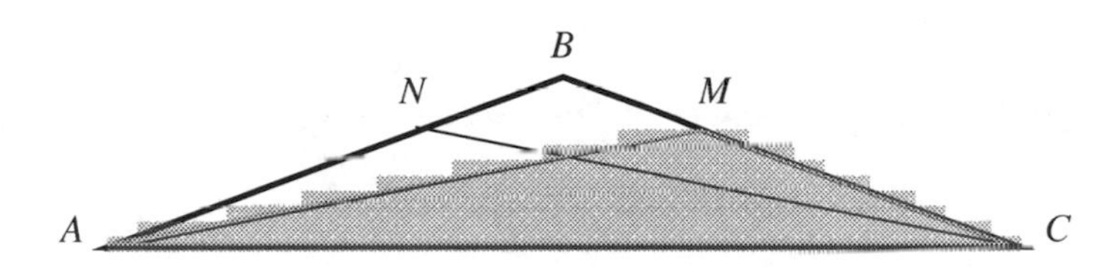

$\angle BAC = \angle BCA$ as the angles opposed the congruent sides in a triangle, and $\angle MAC = \angle NCA$ as halves of congruent angles; hence $\Delta AMC = \Delta CNA$ by ASA. Therefore $AM=CN$ as corresponding sides in these triangles, Q.E.D.

13. This statement is the contrapositive of the statement of Problem 2 from Chapter 4.Review: *If a triangle is isosceles then two altitudes in the triangle are congruent.*

Therefore, the statement follows from the above proof of the hypothesis of Problem 2.

(One may also say that the congruence of two altitudes is necessary upon the congruence of two sides; hence if two altitudes are not congruent, the triangle cannot have two congruent sides.)

16. Let us prove the statement by contradiction: Suppose three points A, B, and C in a plane are non-collinear, and their respective images A', B', C' are collinear, as shown in the diagram below.

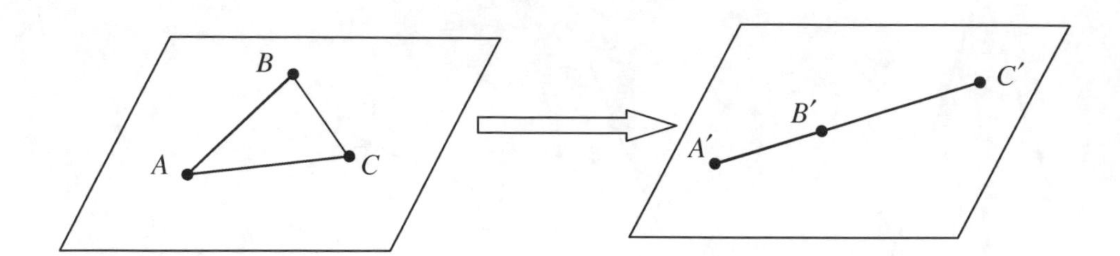

Then, according to the definition of the addition of segments, $A'C' = A'B' + B'C'$. At the same time, according to the triangle inequality, $AC < AB+BC$.

Since under isometries the image of every segment is congruent to the segment, $A'B' = AB$; $B'C' = BC$; $A'C' = AC$, one can obtain from the above equalities and inequality the following result:
$A'C' = A'B' + B'C' = AB + BC > AC = A'C'$, i.e. $A'C' > AC$, which is false. Thus we have arrived to a contradiction; hence the suggestion that an isometry transformed three non-collinear points into three collinear ones has been false, Q.E.D.

Chapter 5

7(c). Construct a triangle, having given two sides and the angle opposite to the least of them.

Solution. Let segments a and b be the given sides, $a < b$, and α be the angle opposed the side equal to a.

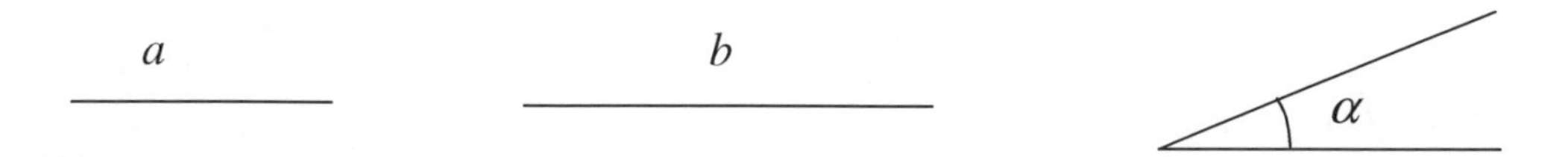

CONSTRUCTION and PROOF. Let us construct an angle congruent to the given angle α with its vertex at some point A (standard construction (a)). Let the sides of this angle be denoted m and l: $\angle A = \angle(l, m)$.

On ray m from point A, lay off a segment $AC = b$ (Compass postulate). From point C describe a circle of radius congruent to a (Compass postulate). If this circle intersects ray l at some point B, then ΔABC is a sought for triangle since in this triangle $BC = a$, $AC = b$, and $\angle A = \alpha$.

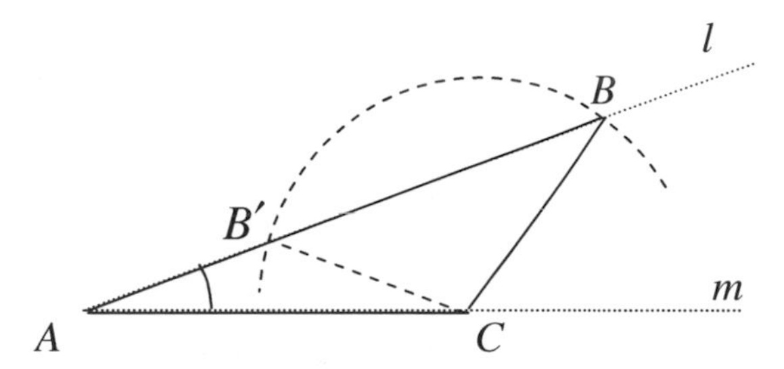

INVESTIGATION. Each step of the construction except finding the point of intersection of the circle with ray *l*, can be carried out and is defined uniquely: any other angle congruent to α can be superimposed on the constructed $\angle A = \alpha$; and by the Compass postulate, there is exactly one such point C lying on a side of the angle that $AC = b$.

According to the Compass Postulate, we can describe from *C* exactly one circle of radius congruent to a. Then one of the following three situations may take place:

(i) The circle intersects ray *l* at two points, B and B', if the distance from *C* to *l* (the perpendicular from *C* onto *l*) is less than a. (Since by the hypothesis $a<b$, point B' cannot fall on the extension of the ray beyond *A*). In this case both ΔABC and $\Delta AB'C$ are solutions; the problem will have exactly two solutions.

(ii) The circle will touch *l* at one point *B* if the distance from *C* to *l* is congruent to a. Then the problem will have exactly one solution.

(iii) The circle will not have common points with *l* if the distance from *C* to *l* is greater than *a*. In this case the problem has no solutions.

10(a). Construct an isosceles triangle having given its altitude (drawn to the base) and a lateral side.

Solution. Let segments *a* and *h* be the given segments: the lateral side and the altitude to the base respectively. $a > h$ since, by the hypothesis, *a* is an oblique and *h* is a perpendicular drawn from the same vertex to the base of the triangle.

a ________________ h ____________

ANALYSIS. Let ΔABC be a sought for triangle with lateral sides $AB = AC = a$, base *BC*, and an altitude to the base $AD = h$.

Then we can view vertex *B* as such a point located on a perpendicular erected to the base, that the distance from *B* to the base is *h*. Also, the distance from each of the other two vertices (*A* and *C*) from *B* is equal to *a*. (See the diagram below).

Based on this consideration, we can propose the following plan for solving the problem: we shall suggest that the base of a required triangle lies on some line and erect a perpendicular equal to *h* to this line. The "upper" endpoint of this perpendicular (the endpoint that is exterior to the line) will be the vertex (*B*) opposed to the base. Then we shall find on the line of the base two other vertices: these will be the points that are located at the distance a from B.

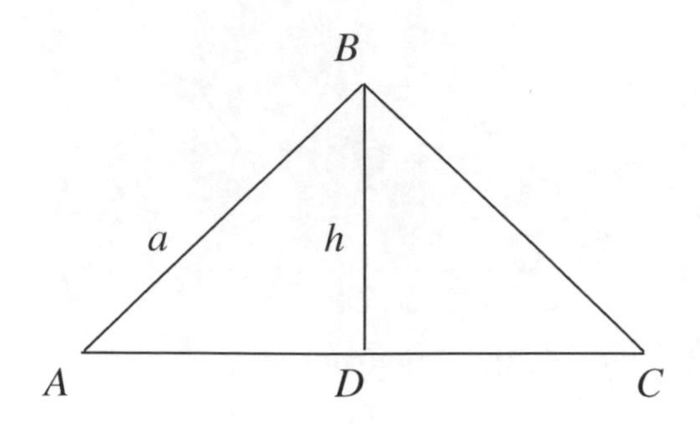

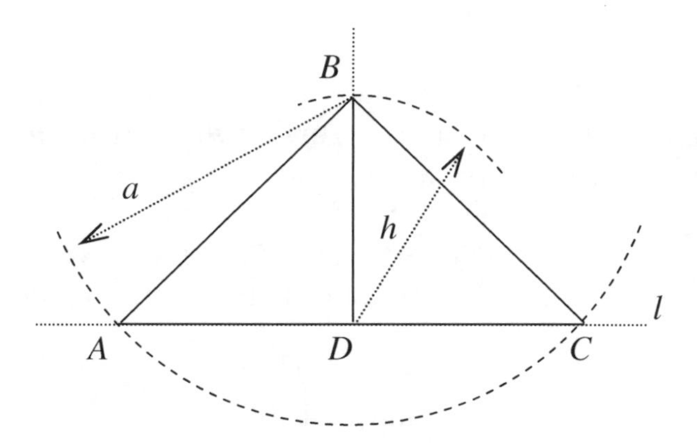

CONSTRUCTION. Let l be some straight line.
From some point D on l, erect a perpendicular to l (construction (g)).
From D, lay off on this perpendicular a segment $DB = h$.
From B, describe a circle of radius a (Compass Postulate). Since $a > h$, the circle will intersect l at two points; let us name them A and C.
Thus we have constructed ΔABC. Let us prove that this is a sought for triangle.

PROOF. In ΔABC, the lateral sides $BA = BC = a$ by construction, as two radii of the same circle; The altitude from vertex B is $BD = h$ by construction. Hence, ΔABC does satisfy the required conditions, Q.E.D.

INVESTIGATION. Let us prove that a triangle satisfying the given conditions is *unique* in the following sense: any other triangle that satisfies these conditions will be congruent to ΔABC we have constructed.

Suppose ΔMNP is another triangle satisfying the given conditions: $NM = NP = a$, and $NE = h$, where ME is the altitude drawn to the base MP.

Then $\Delta MNE = \Delta ABD$ by the hypotenuse and a leg; similarly, $\Delta PNE = \Delta CBD$. Therefore, $\Delta MNP = \Delta ABC$ (it is easy to prove, for example, by superimposing the triangles: ME onto BD, or in some other way, e.g. by *SSS*), Q.E.D.

10(c). Construct an isosceles triangle having given its base and an altitude drawn to a lateral side.

Solution. Let segments b and h be the given segments: the base of a sought for triangle is required to be congruent to b whereas an altitude to a lateral side must be congruent to h. Let us notice that $b > h$ (Why?).

b h

ANALYSIS. Let ABC be a sought for triangle, i.e. $AC = b$, $AB = BC$, and $AD = h$, where $AD \perp BC$.

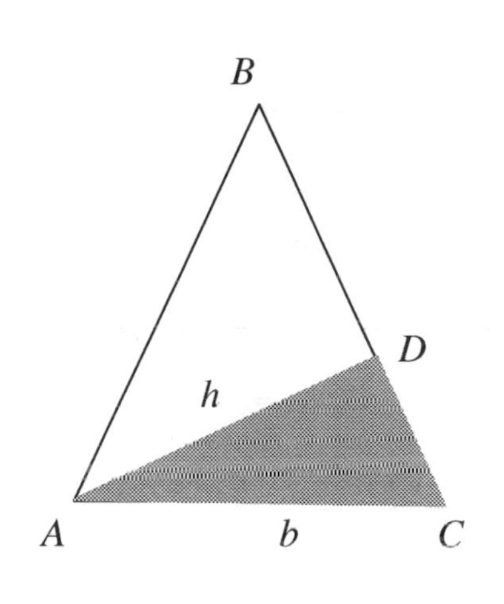

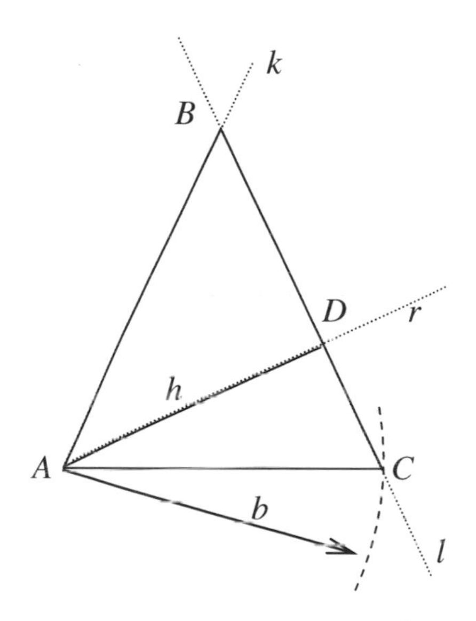

ΔADC is right, and we can construct such a triangle by its hypotenuse $AC = b$ and a leg $AD = h$. Then we shall use this triangle as an auxiliary figure in order to construct ΔABC.

CONSTRUCTION. Let us draw an arbitrary ray r emanating from some point A.
From A, lay off on this ray a segment $AD = h$.
Through D, draw a line l perpendicular to r.
From A, describe a circle of radius b (Compass Postulate); since $b > h$, it will cut line l at two points. Let one of these points be denoted C.
Join A with C. (This completes the construction of ΔADC, formed by the base and an altitude to a lateral side in a sought for triangle).
From A as a vertex, draw a ray k that lies on the same side of AC as point D and makes with AC an angle congruent to $\angle DCA$. If this ray intersects line l at some point B, this will complete the construction of ΔABC with $\angle ABC = \angle ACB$.
Let us prove that this is a sought for triangle.

PROOF. ΔABC is isosceles, since by construction $\angle ABC = \angle ACB$ (Th. 4.6.2), which implies $AB = BC$. By construction, the base $AC = b$, and AD, the altitude to a lateral side, is congruent to h. Thus ΔABC is an isosceles triangle with the given base and altitude to a lateral side, Q.E.D.

INVESTIGATION. Any other triangle that satisfies the same conditions will be congruent to ΔABC, which follows from the congruence of the right triangles formed by the given bases and altitudes to lateral sides (the rest is easy to prove by superposition).

In this sense the solution *is unique if it exists.* (Surprisingly enough we cannot prove the existence of a solution for any given set of b and h such that $b > h$, since we are not sure whether ray k will intersect line l. In order to guarantee that they will intersect we would have to assume one more postulate, the so-called Parallel Postulate, which will be introduced in Chapter 6).

19b. Construct a triangle having given its base, an angle adjacent to the base, and the difference of the remaining (lateral) sides.

Solution. Let segment b be the base, angle α – one of the angles adjacent to the base, and segment d – the difference between the lateral sides.

ANALYSIS. Suppose ΔABC is a sought for triangle with the base $AC = b$, $\angle BAC = \alpha$, and the difference between the remaining sides congruent to d.

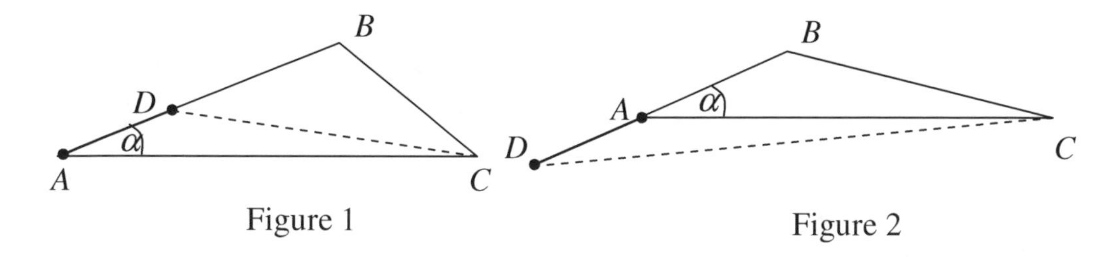

Figure 1 Figure 2

There are two possible cases, illustrated in the above diagrams: angle α may be opposed (i) the least of the lateral sides (Figure 1) or
(ii) the greatest of the lateral sides (Figure 2).
In case (i), lay off from B a segment $BD = BC$ on BA, the greater of the lateral sides. Then in ΔADC: $AC = b$, $\angle DAC = \alpha$, and $AD = d$; therefore we can construct ΔADC by SAS.

In case (ii), when BC is the greater of the lateral sides, lay off from B on the ray BA, a segment $BD = BC$.
Then in ΔADC: $AC = b$, $\angle DAC$ is supplementary to α, and $AD = d$; therefore we can construct ΔADC by SAS.

In any of these cases (i or ii), ΔDBC is isosceles with *BD*=*BC* (by construction), hence the altitude drawn from *B* to *CD* will be a perpendicular bisector of *CD*. Then, if ΔADC is constructed, we can find vertex *B*, opposed to the base *AC*, as the point of intersection of the line extending *AD* with the perpendicular bisector of *CD*.

CONSTRUCTION. In any of the considered cases, construct ΔADC by SAS, and extend *AD* into a straight line till it intersects the perpendicular bisector of *CD*. Let us name the point of intersection *B*. Then ΔABC is a sought for triangle, which is easy to prove (the proof follows).

PROOF. In any of the considered cases (i or ii), in ΔABC: $AC = b$, $\angle BAC = \alpha$, and the difference between *AB* and *BC* (i) or *BC* and *AB* (ii) is congruent to d; therefore ΔABC is a sought for triangle, Q.E.D.

INVESTIGATION. ΔADC is defined uniquely by *SAS* in each of the cases (i or ii). The point of intersection of the perpendicular bisector of *CD* (there is only one such perpendicular bisector, - why?) with the line containing *AD* is also unique (AXIOM 1), if it exists, of course.
Therefore, if that point of intersection exists, the problem has exactly two solutions, corresponding to the cases (i) and (ii) respectively.
Remark. This problem of constructing a triangle satisfying the given conditions, is solved in the so-called *neutral geometry*, Euclidean geometry based on the first four postulates. In order to complete the investigation and show that the perpendicular bisector of *CD* intersects with the line containing *AD*, we need one more Euclidean Postulate, called the *parallel postulate*. It follows from that postulate that an oblique and a perpendicular to the same line will necessarily intersect.

23. Such a place for the station will be located at the point of intersection of the railroad with the perpendicular bisector of *AB*, if of course, these two intersect (see the remark at the end of Chapter 5 and the remark to the solution of problem 19b).

Sections 6.1-3

22. It should be noticed that this is a problem of constructing a triangle by SAA condition in the particular case when the angle opposed the given side is right.

Let the segment *c* and angle α in the diagram below be the given hypotenuse and an acute angle of a sought for right triangle.

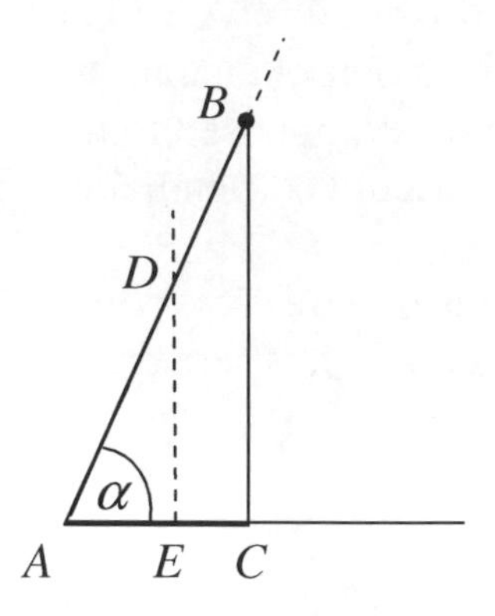

CONSTRUCTION and SYNTHESIS (PROOF).

Let us start with constructing a right triangle with an acute angle congruent to the given angle α (its hypotenuse is not necessarily congruent to *c* yet) . In order to do this, we construct an angle congruent to α with its vertex at some point *A* and erect a perpendicular to one side of this angle from an arbitrary point *E* on that side.

As it follows from the parallel postulate (Th.6.3.3), this perpendicular will intersect the other side of the angle at some point *D*. If, by a lucky coincidence, $AD = c$, then $\triangle ADE$ is a sought for triangle, and the construction is completed.

If $AD \neq c$, we shall lay of on the other side of the angle (the side that does not contain *E*) a segment $AB = c$ and through C we shall draw a line parallel to *DE*. This line will intersect the ray *AE* at some point *C*, and it will be perpendicular to *AE* (why?).

Then $\triangle ABC$ will be a sought for triangle since $\angle BCA$ is right by construction (the corresponding angles formed two parallel lines, *BC* and *DE*, with the ray *AE* are congruent), $\angle A = \alpha$, and $AB = c$ by construction.

INVESTIGATION.

By SAA condition, any other right triangle with the same hypotenuse and an acute angle, will be congruent to the one constructed above.

Remark. This construction can be carried out in *neutral geometry*: in the last step, instead of drawing a line parallel to *DE* through *B*, we could drop a perpendicular from *B* onto *AE*. (Of course, we can draw a line parallel to a given line in neutral geometry, but we cannot conclude whether we have obtained the right angle at the end). This option (of dropping a perpendicular) does not exist in a general SAA construction, when the angle opposed to the side *AB* is not right.

Section 6.6

7. Given a point in the interior of an angle, draw through this point a segment with the endpoints on the arms of the angle and such that the segment is bisected by the given point. (In other words: draw a *chord of the angle* that is bisected by the given point).

ANALYSIS. Let angle EAF be the given angle with point P in its interior. Suppose points B and C on the sides of the angle are such points that $BP = PC$. Then, if we construct a parallelogram $ABCD$ with three vertices lying at B, A, and C, segment BP will be a diagonal in this parallelogram Hence AD will be another diagonal, and it must be bisected at point P (Theorem 7.1.3).

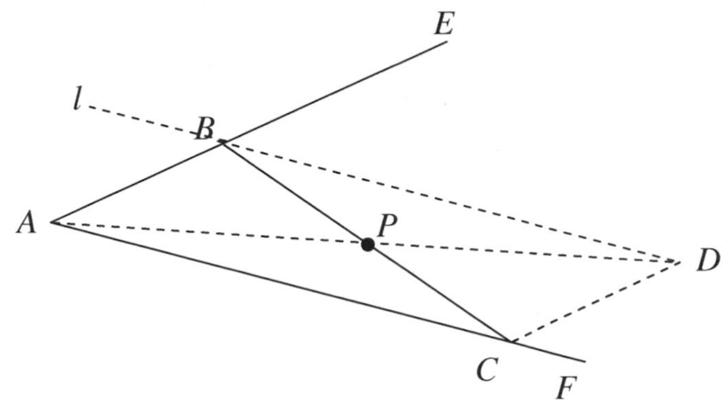

CONSTRUCTION. Through the given point P in the interior of the given angle BAC, draw a ray and lay off on this ray from point P segment $PD=AP$.

Through point D draw lines parallel to the arms of the given angle. The line l parallel to AF will intersect AE at some point B (this line cannot be parallel to AE since it is parallel to AF, and by AXIOM5, there is only one line parallel to a given line (l) through a given point (A). Similarly, the line parallel to AE will cut AF at some point C.

Then $ABCD$ is a parallelogram (by construction, and P is the point of intersection of the diagonals, therefore BC is a diagonal of a parallelogram, and $BP=PC$ (Theorem 7.1.3). Thus BC is the sought for segment.

PROOF. The proof is contained in the last paragraph of the CONSTRUCTION.

INVESTIGATION. Every step of the above construction is justified by the axioms and theorems of Euclidean geometry, and we did not make any special assumptions concerning the given angle and point in its interior; hence one solution of the problem does exist. Let us show that there is no other solutions. We shall prove this by contradiction (RAA).

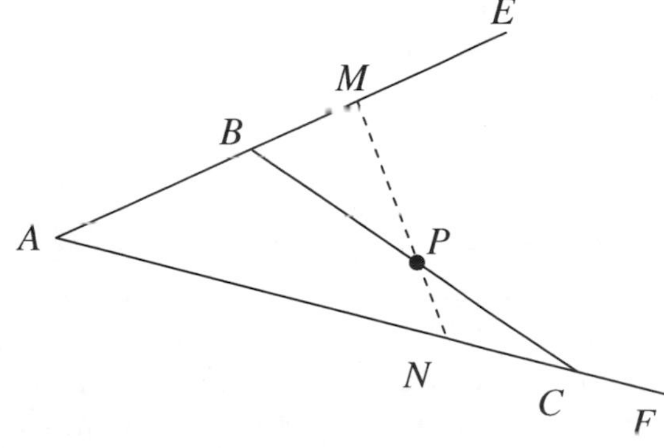

Suppose MN is another segment bisected at P: $MP=PN$. Then $\triangle NPC = \triangle MPB$ by SAS: $BP = PC(suggestion)$; $MP = PN(construction)$; $\angle BPM = \angle CPN(vericals)$. Then $\angle PNC = \angle PMB$ as corresponding angles in congruent triangles, but this contradicts

Theorem 6.3.1, since the sides AE and AF of the given angle are not parallel.

Hence the problem has a unique solution.

Section 7.1

32. Let segments b and c be the given sides and m be the median included between them, i.e. the median bisecting the third side.

b c m

ANALYSIS.

Suppose $\triangle ABC$ is a sought for triangle with $AC = b, AB = c, AM = m$.

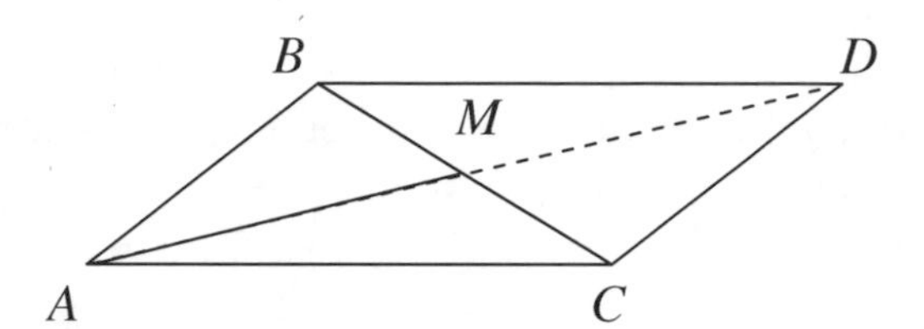

Let us extended ray AM beyond M by a segment $MD=AM$. Join D with B and C. It is easy to see that $\triangle ABM = \triangle DCM$ by SAS condition ($BM=MC$ since AM is a median, $\angle AMB = \angle DMC$ as verticals, and MD is constructed to be congruent to AM).

Similarly, $\triangle AMC = \triangle DMB$.

Then the whole figure (parallelogram $ABCD$, whose diagonals intersect at M), is symmetric in M and consists of two congruent triangles, ABC and DCB, which are symmetric to each other in M.

Also this quadrilateral consists of two congruent and symmetric in M triangles ACD and ABD. Each of the latter triangles can be constructed by SSS condition: $AC = b, CD = AB = c$, and $AD = 2m$.

After one of this triangle is constructed, it can be completed into $ABCD$ by adding to it its symmetric in M image. Then, by drawing diagonal BC we obtain two symmetric each other in M triangles, each of them a sought for triangle.

CONSTRUCTION.

Construct $\triangle ADC$ by SSS condition, with $AC = b, CD = AB = c$, and $AD = 2m$.

From D, describe a circle of radius congruent to $AC=b$, and from A describe a circle of radius congruent to $CD=c$. The circles will intersect at two points; let the one lying on the opposite of C side of AD be called B. (Alternative way: through D, draw a line parallel to AC, and through A draw a line parallel to CD; they will intersect at some point B creating a triangle ABD, which can be shown to be symmetric to $\triangle DCA$ about M, the point where AD cuts BC.)

Draw BC. Thus we have obtained $\triangle ABC$. Let us show that this is a sought for triangle.

PROOF.

By construction, $DB = b = AC$, $AB = c = CD$, and AD is a common side of $\triangle ABD$ and $\triangle DCA$; hence these triangles are congruent by SSS condition. Then $ABDC$ is a parallelogram, since $\angle DAC = \angle ADB$ as corresponding angles in congruent triangles, and hence $DB \parallel AC$ and $DB=AC$ by construction.

Let M be the point of intersection of the diagonals of this parallelogram. Then, according to Th.7.1.3, $AM = MD = \frac{1}{2}2m = m$, and $CM=MB$, which means that AM bisects BC. Also, $AC=b$, and $AB=c$ by construction. Therefore $\triangle ABC$ is a sought for triangle, Q.E.D.

INVESTIGATION.

In the above construction, $\triangle ADC$ is defined uniquely by SSS condition, and it has exactly one symmetric image about M, the midpoint of AD. Hence, the parallelogram is defined uniquely, and so is $\triangle ABC$, since there is only one line through B and C (Axiom1).

Section 7.2

14. Let a and b, and be the given sides, and h be the altitude drawn to b.

ANALYSIS.

Suppose $\triangle ABC$ is a sought for triangle with $AC = b, CB = a$, and its altitude BD is congruent to h.

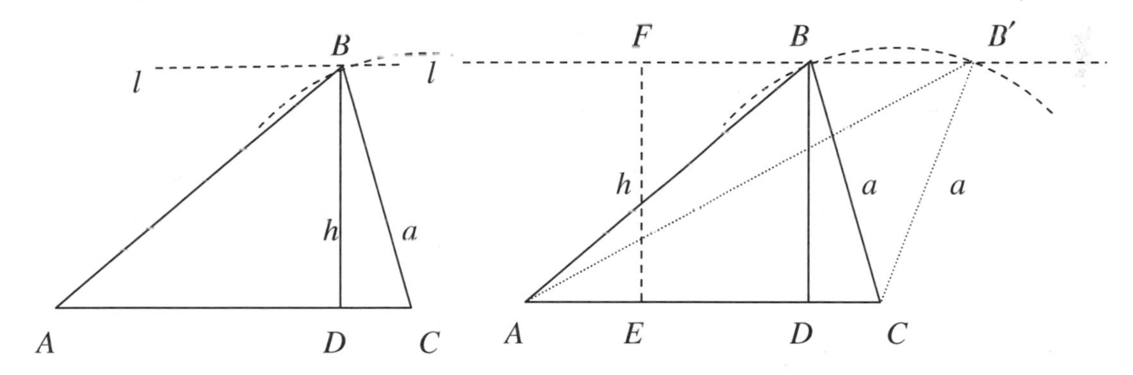

Vertex B of the triangle stands at the distance h from the line containing the opposite side AC; also the distance from the vertex B to one of the other two vertices, e.g. vertex C, is congruent to the given segment a. Thus, vertex B can be viewed as a point of intersection of two loci: a line standing distance h from AC and a circle of radius a centered at C.

CONSTRUCTION.

On an arbitrary line, lay off a segment $AC = b$. From an arbitrary point E of this line, erect a perpenicular and lay off on that perpendicular a segment $EF = h$. Through F, draw a line l parallel to AC. From C, describe a circle of radius a. If it intersects line l at some point B, join B with A and C to obtain $\triangle ABC$.

Let us prove that the triangle obtained as a result of this construction is a sought for triangle.

PROOF.

In $\triangle ABC$, $AC = b$ and $CB = a$ by construction. Let *BD* be the altitude dropped from *B* onto *AC*. Since $BD \perp AC$, it will be perpendicular to *l* as well (why?). Then *EFBD* is a rectangle, and hence $BD = EF = h$. Thus, $\triangle ABC$ satisfies all the given conditions, Q.E.D.

INVESTIGATION.

There are two lines lying at the distance h from *AC*, one in each half-plane. A construction performed for one of these lines, can be carried out for the other one and it will result in a symmetric in *AC* and therefore congruent figure. In this sense the first step may be deemed as uniquely defined.

The circle of radius a :

(i) will not intersect line *l* if $a < h$, and there is no solution in this case (a side in a triangle cannot be less than an altitude to some other side as an oblique cannot be less than a perpendicular to the same line);

(ii) will intersect line *l* at two points, *B* and B' creating two different solutions: triangles ABC and $AB'C$ if $a > h$ (this is a so-called *generic* (most general, or non-special case);

(iii) will touch line *l* at exactly one point if $a = h$. In this case $\triangle ABC$ will be right.

18. Let *b* be the given side, *h* be the altitude drawn to *b*, and *m* be the median drawn to some other side of the sought for triangle.

ANALYSIS.

Let *ABC* be a sought for triangle with $AC = b$, $BD = h$ the altitude to *AC*, and $CM = m$ the median bisecting *AB*.

Extend *CM* by a segment *MN*=*CM* and join *N* with *A* and *B*. *ANBC* is a parallelogram (Th. 7.1.4) that can be constructed since it consists of two congruent triangles *ANC* and *BCN* (each of them can be constructed by the method described in the solution for problem 14 of this section).

Then, a sought for triangle can be obtained by means of drawing the second diagonal of that parallelogram.

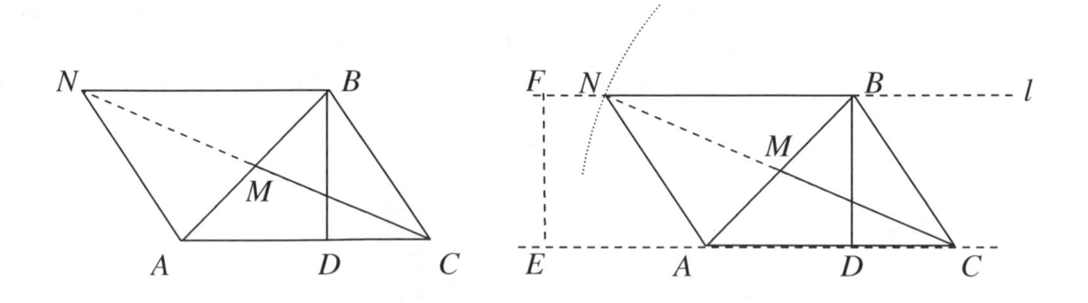

CONSTRUCTION.

On an arbitrary line, lay off a segment $AC = b$. From an arbitrary point E of this line, erect a perpenicular and lay off on that perpendicular a segment $EF = h$. Through F, draw a line l parallel to AC.

From C, describe a circle of radius $2m$. If it intersects line l at some point N, join N with A and through C draw a line parallel to AN to obtain a parallelogram $ANBC$. Draw AB to obtain $\triangle ABC$. Let us prove that it is a sought for triangle.

PROOF.

By construction, $AC = b$, and the distance between any point of l to the line containing AC is congruent to h, hence the altitude $BD = h$. By construction, $CN = 2m$, and since the diagonals of a parallelogram bisect each other, $CM = \frac{1}{2}CN = m$ is the median drawn to AB.

Thus, $\triangle ABC$ does satisfy all of the given conditions.

INVESTIGATION.

The above construction can be carried out in either half-plane formed by the line containing AC. In each half-plane there is exactly one line (why?) parallel to AC and standing at distance h from it. For each figure constructed in the "upper" half-plane there will be a symmetric in AC and therefore congruent figure in the "lower" half-plane.

Now let us discuss possible numbers of solutions in one half plane.

Line l is defined uniquely. The next step of the construction involved finding the points of intersection of that line with the circle of radius $2m$ described from C.

If $2m > h$, i.e. the distance from C to the line l is less than the radius of the circle, part of the line will lie in the interior of the circle, and there will be two points of intersection, N and N'. For each of this points (see the diagram below) the construction can be completed and will result in a triangle ABC or $AB'C$ respectively. Thus, if $2m > h$, the problem has exactly two solutions.

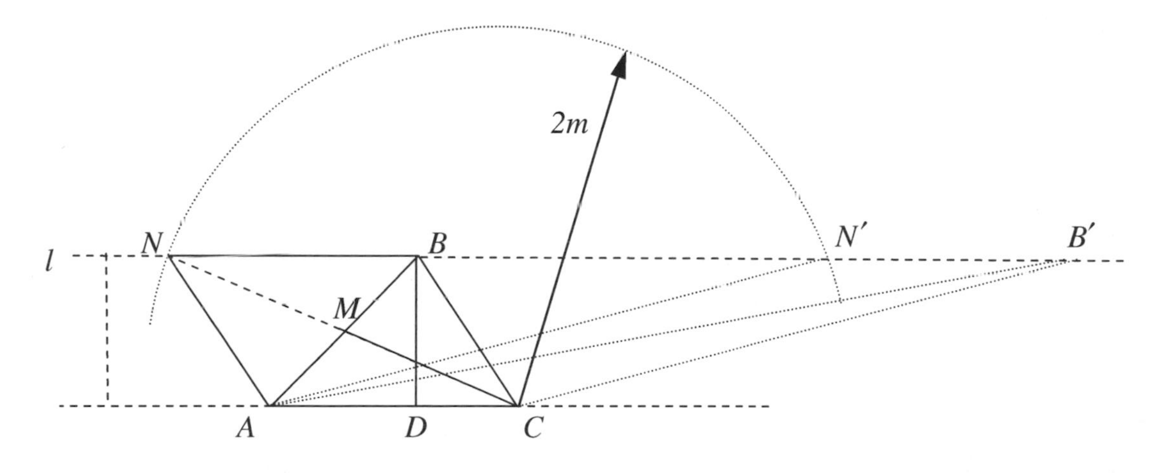

If $2m = h$ (this is possible only if the median *AM* is perpendicular to *AC*), there will be exactly one point *N* at which the circle of radius $2m$ touches line *l*, and therefore there will be exactly one solution in this case (see the diagram below).

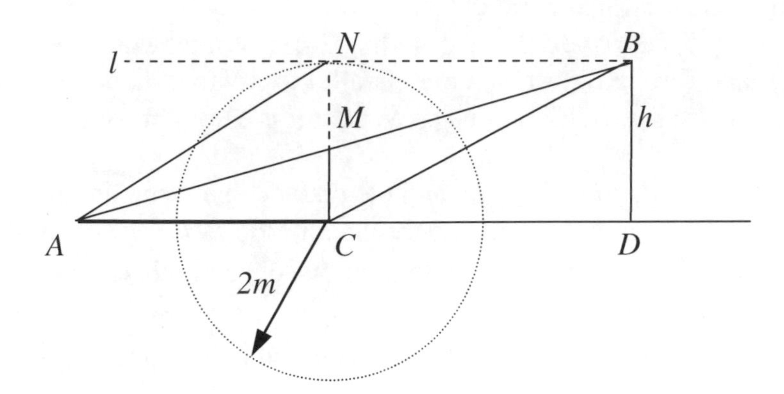

$2m$ cannot be less than *h*, since *h* is the shortest distance between the points of the two parallel lines (*AC* and *l*). Thus all the possible cases are exhausted.

Section 7.4

2.

(i) May one of the angles that include the greater base of a trapezoid be obtuse?

Solution. Let us show that there exists a trapezoid in which one of the angles including the greater base is obtuse.

Let *ABCD* be a parallelogram with $\angle ADC$ obtuse. Extend *AD* beyond *D*. From *C*, drop a perpendicular onto the line containing *AD*. Let the foot of the perpendicular be called *E*. *E* falls on the extension of *AD* beyond *D* since otherwise we would obtain a triangle *ECD* with one interior angle ($\angle CED$) right and another interior angle ($\angle EDC$) obtuse (see the left diagram below).

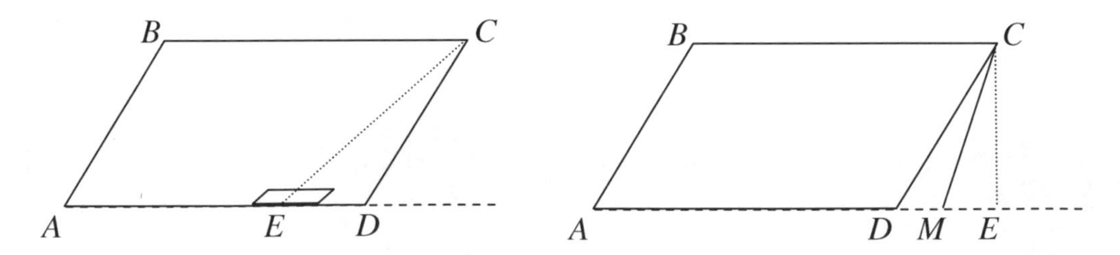

Let *M* be an *interior* point of the segment *DE* (a point is called an *interior* point of a segment if it lies on the segment and does not coincide with any of its endpoints). Then *ABCM* is a trapezoid with the bases *AM* and *BC*.

$AM > BC$, since *ABCD* is a parallelogram, in which $AD = BC$, and therefore $AM = AD + DM > AD = BC$.

Also, $\angle DMC$ is obtuse since it is exterior in $\triangle CME$, where $\angle MEC$ is right.

Thus, *ABCM* is a trapezoid in which one of the angles including the greater base is obtuse.

(ii) May both angles that include the greater base of a trapezoid be obtuse?

Solution. Let us show that a trapezoid with two obtuse angles including the greater base cannot exist.

Suppose *ABCD* is a trapezoid with the bases *BC* and *AD*, $AD > BC$, and each of the angles *BAD* and *ADC* obtuse.

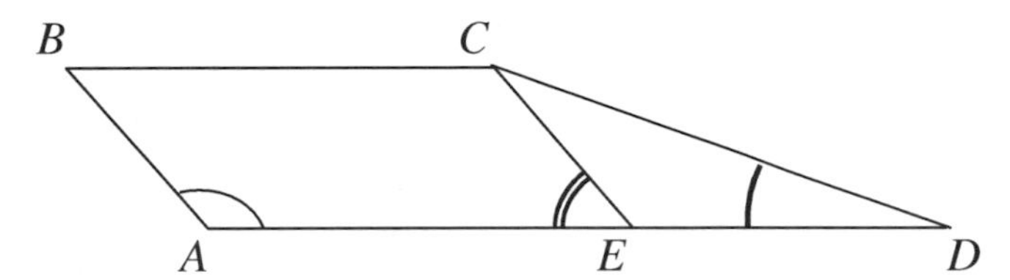

On *AD*, lay off a segment *AE* congruent to *BC*. Since $AD > BC$, and *AE*=*BC*, we conclude that *AE* is less than *AD*, and hence point *E* lies in the interior of *AD*.

Since $AE \parallel BC$ and *AE*=*BC*, ABCE is a parallelogram, according to Th. 7.1.2. Then, by Th. 7.1.1, $\angle CEA$ is supplementary to $\angle BAD$, which is obtuse, and therefore $\angle CEA$ must be acute. On the other hand, $\angle CEA$ is exterior to ΔCED, and hence it must be greater than $\angle CDE$, which is obtuse by our assumption.

Thus we have arrived to a contradiction, which proves that a trapezoid with the greater base included between two obtuse angles cannot exist, Q.E.D.

24. Let the given segments *a*, *p*, *q*, and *d* be respectively a base, the lateral sides and a diagonal of a sought for trapezoid.

a p q d

ANALYSIS.

Let *ABCD* in the left diagram below be a sought for trapezoid with $AD = a, AB = p, CD = q,$ and $AC = d$. Thus we have assumed that the given diagonal forms a triangle with the given base and the side congruent to *q*.

We can construct ΔACD by SSS condition. Point *B*, the fourth vertex of the trapezoid, can be found as a point of intersection of a line passing through *C* parallel to *AD* and an arc of radius *p* described from *A*.

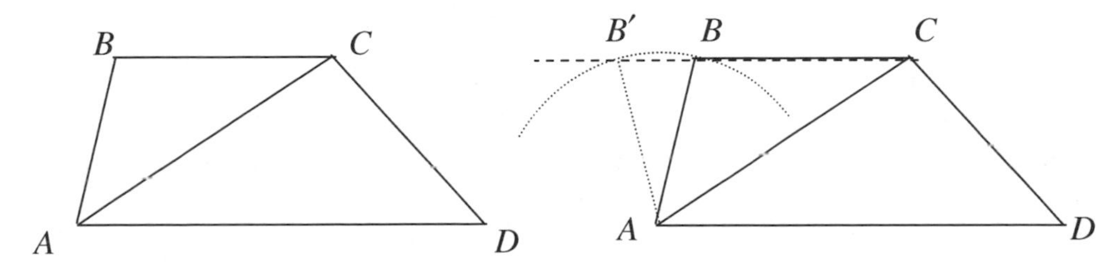

CONSTRUCTION and SYNTHESIS (PROOF)

Construct (if possible) by SSS condition a triangle ACD with $AC=d$, $CD=q$, and $AD=a$. Then, through C draw a line parallel to AD, and from A describe an arc of radius p. If they intersect at some point B, this point will be the fourth vertex of a sought for trapezoid. This follows from the construction: it is a trapezoid since CB has been drawn parallel to AD, and in this trapezoid $AD = a, AB = p, CD = q$, and $AC = d$.

INVESTIGATION.

The above construction has been based on the suggestion that the given diagonal and the lateral side congruent to q emanate from the same endpoint (A) of the given base. Thus, everything in this construction and the following investigation should be repeated when the sides p and q are interchanged. Such an interchange will create solutions that differ from the ones obtained by the above procedure.

ΔACD can be constructed if and only if the triangle inequalities are satisfied for a, d, and q. Also, one can start If this triangle can be constructed, the trapezoid will exist iff the other lateral side (in this case p) is greater than or congruent to the altitude h of ΔACD dropped onto AC. If $p>h$, the problem will have two distinct solutions ($ABCD$ and $AB'CD$ in the right diagram above); if $p=h$, the arc will touch the line of the "upper" base only at one point, and there will be exactly one solution.

The consideration of the cases created by interchanging p and q may double the number of solution, although it may happen that no new solutions will be generated (if, for instance, $p+d<a$, and thus a triangle with such sides does not exist).

25. Let the segments l, m, n, and d in the diagram below be the given difference between the bases, the lateral sides and a diagonal of a sought for trapezoid respectively.

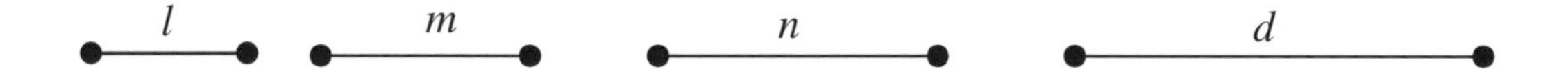

ANALYSIS. Suppose the trapezoid in the figure below is a sought for trapezoid:

$AD \| BC$; $AD - BC = l$; $AB = m$; $CD = n$; $AC = d$.

Let us translate AB parallel to itself so that B moves into C (see the diagram below). Thus we obtain a parallelogram $ABCE$ with $AE=BC$ and a triangle CED, in which:

$CE=AB= m$ (AB is translated into CE),

$ED = AD - AE = (\text{since } AE = BC) = AD - BC = l$, and $CD = n$.

Then we can construct ΔCED by SSS.

If we have this triangle constructed, we can translate CE back, so that C falls on B. The location of B can be determined as a point that stands at the distance d from point D on the line that passes through C parallel to ED.

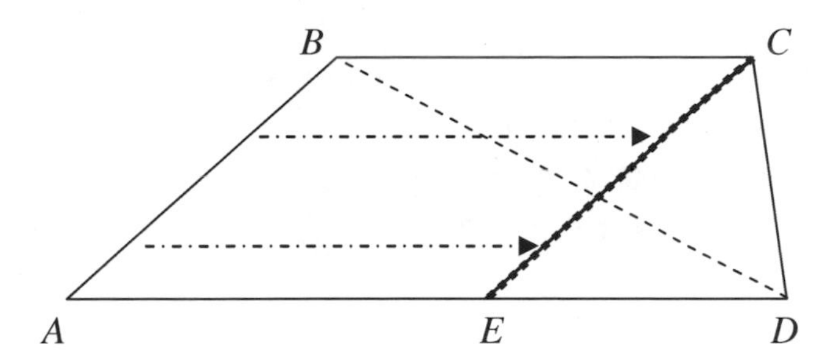

CONSTRUCTION. Construct ΔCED with $ED = l$, $EC = m$, and $CD = n$.

Through *C* draw a line *k* parallel to *ED*. From *D* as a centre, describe an arc of radius *d*. This arc will intersect line *k* at two points, *B* and *F*.

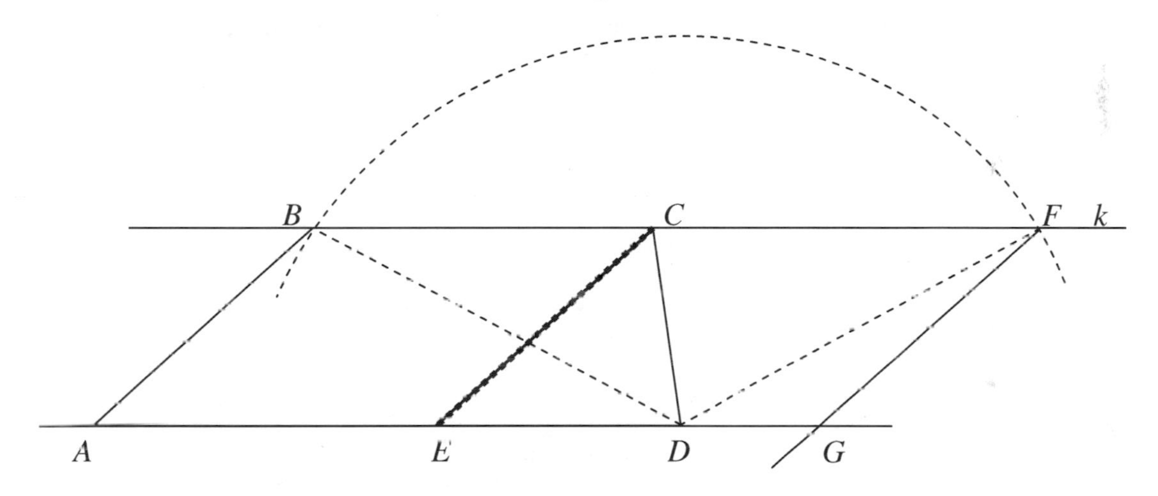

Through each of these points draw a line parallel to *CE*. Each of these two lines will intersect (why?) the line passing through *E* and *D* (extension of *ED*) at some points *A* and *G* respectively.

Thus we obtain two trapezoids: *ABCD* and *ECFG*. Each of them satisfies the conditions of the problem, and hence each of them is a sought for trapezoid.

PRROF. The proof follows directly from the construction.

In *ABCD*: *ABCE* is a parallelogram by construction, hence *AB* = *EC* as opposite sides of a parallelogram, and therefore *AB* = *m*. *CD* = *n* by construction. *BC* is parallel to *AD* by construction, and hence *ABCD* is a trapezoid. Since *BC* = *AE* as opposite sides of a parallelogram, it follows that $AD - BC = AD - AE = l$. Diagonal $DB = d$ by construction. Therefore, all the conditions are satisfied, and *ABCD* is a sought for trapezoid.

Similarly one can prove that *ECFG* is a sought for trapezoid.

INVESTIGATION. Our construction procedure leads to exactly two solutions. This does not prove yet that other solutions cannot be constructed, maybe by means of some other procedure.

In order to prove that only these two solutions are possible, one should prove that any other trapezoid with the given data will be congruent to one of our solutions.

Section 7.5

2. Let the segments *a*, *b*, and *c* be three consecutive sides (they follow in this order) of a required quadrilateral, and angles α and β are the angles enclosing the unknown side of this quadrilateral: α is the angle between the unknown side and the side congruent to *a*, and β is the angle between the unknown side and the one congruent to *c*.

a *b* *c* α β

ANALYSIS.

Suppose *ABCD* in the diagram below is a sought for quadrilateral with $AB = a$, $BC = b$, $CD = c$, $\angle DAB = \alpha$, and $\angle ADC = \beta$.

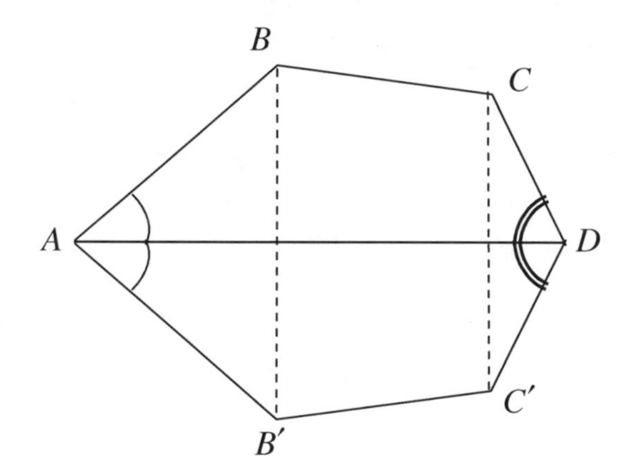

Reflect *B* and *C* in the unknown side *AD*. Let B' and C' be their respective images. The obtained figure $ABCDC'B'$ consists of two isosceles triangles: BAB' and CAC', and an isosceles trapezoid $BCC'B'$.

The triangles can be constructed by SAS conditions. The bases of these triangles are also the bases of the trapezoid, and the latter can be constructed by its sides. Then we can construct $ABCDC'B'$ and by drawing the diagonal AD we shall obtain a sought for quadrilateral.

CONSTRUCTION and SYNTHESIS.

Construct by SAS conditions two isosceles triangles: $\Delta BAB'$ with the angle 2α between two sides, each congruent to *a*, and $\Delta CDC'$ with the angle 2β between the sides congruent to *c*. Let us denote: $BB' = p$; $CC' = q$.

Construct an isosceles trapezoid with the bases *p* and *q* and each lateral side equal to *b*. This can be done as follows (see the diagram below):

Construct an isosceles triangle *BEM* with the lateral sides *b* and *b*, and its base $BM = p - q$ lying on BB' if $p > q$ (otherwise, the third side should be $q - p$ and the base CD of the triangle will lie on CC'): From *B*, lay off on BB' a segment *BM*

$= p - q$; then from *B* and from *M* describe arcs of radii *b*. They will intersect iff $2b > p - q$. Let us name their point of intersection *E*. Translate *EM* parallel to itself along the extension of *BM* through a segment congruent to *q* (the least of *p*, *q*). The segment obtained by translating *EM* will have B' as an endpoint (since $BM = p - q$, and then $MB' = q$). Let us name another endpoint *F*.

We can prove that $BEFB'$ is a sought for trapezoid: it is a trapezoid since *EF* is parallel to *BM* by construction and its bases are *p* and *q* by construction; also by construction one of the lateral sides, *BE*, is congruent to *b* whereas the other one, $B'F$, is obtained by means of a parallel translation from *EM*, which is congruent to *b* by construction.

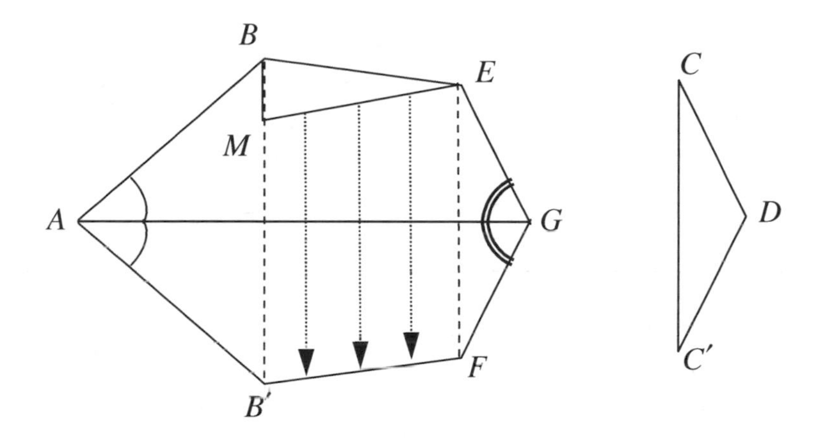

To complete the construction of figure $ABCDC'B'$, we construct a triangle *EFG* congruent to $\Delta CC'D$ with $EF = q = CC'$ as a base. This can be done by describing the arcs of radius *c* from *E* and from *F*; iff $2c > q$, they will intersect at some point *G*.

By construction, the obtained figure $ABGEFGFB'$ consists of two attached to each other sought for quadrilaterals with *AG* as their common side. Then draw *AG* to obtain a sought for quadrilateral (and its reflection image in *AG*).

INVESTIGATION.

As we have already mentioned, the above construction is to be carried out when $p > q$. If $p < q$, a similar construction of the trapezoid (with BB' as the least base) can be carried out. In case $p = q$, the trapezoid will degenerate into a rectangle, which can be constructed by erecting from *B* and B' perpendiculars congruent to *b* and joining their endpoints.

Also, the above method describes the construction for the case when both α and β are acute angles. There are four possible versions of the described method of solving the problem that correspond to various ways of making $ABGEFGFB'$ out of two triangles and a trapezoid. They are presented in the diagram below (in each instance α is denoted by a single arc and β by a double arc, and the sought for quadrilateral is shown in bold continuous lines):

(a) both α and β are acute; (b) α is acute and β is obtuse;
(c) both α and β are obtuse; (d) α is obtuse and β is acute.

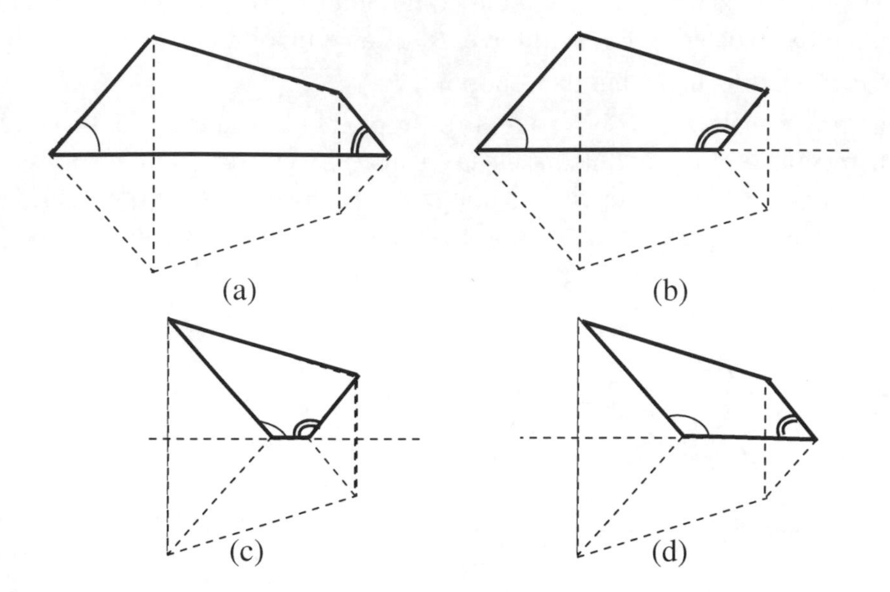

(a) (b) (c) (d)

Each step of the described (or analogous) procedure is defined uniquely if it can be performed. Triangles ABB' and DCC' can always be constructed by SAS condition, whereas an auxiliary triangle (ΔMBE in the above description) is constructed by SSS condition, which is possible iff the triangle inequalities hold.

In some cases (see, for instance, the left and right figures at the diagram below) a second solution leading to a construction of a quadrilateral with intersecting sides is possible. Such a quadrilateral is usually called *complete* as opposed to a quadrilateral with non-intersecting sides, which is called *simple*. Usually complete quadrilaterals are not considered as quadrilaterals in elementary geometry courses.

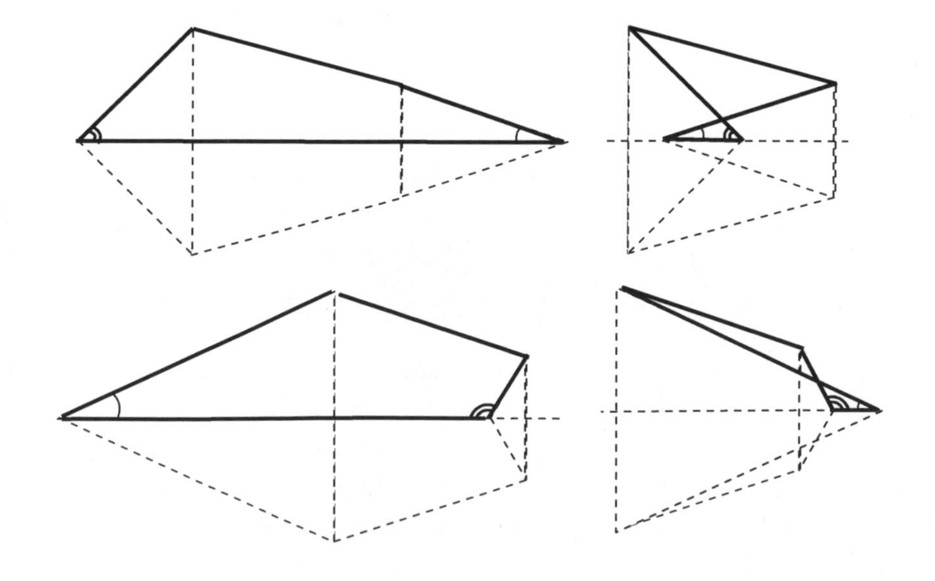

4. Let the segments a, b, c, and d in the diagram below be the given sides of a sough for quadrilateral. We can suggest without loss of generality that they are consecutive sides following in the order listed above, and the angle between the sides a and d is the one that is bisected by a diagonal.

ANALYSIS. Suppose $ABCD$ in the figure below is a sought for quadrilateral with $AB = a,\ BC = b,\ CD = c,\ DA = d,$ and $\angle BAC = \angle DAC$.

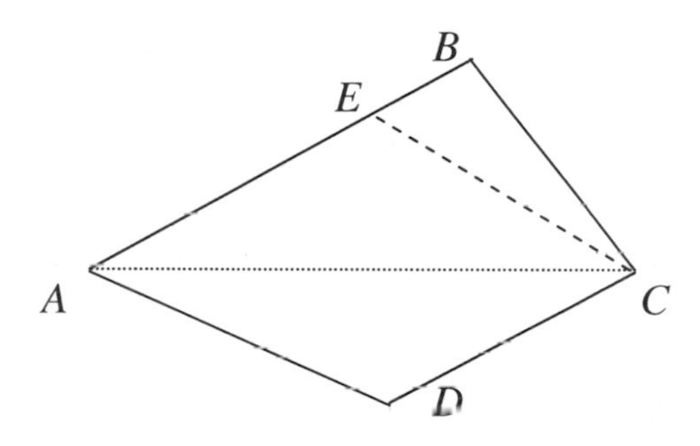

If we reflect AD in AC, its image AE will fall on AB since $\angle DAC = \angle BAC$ (we suggest in the diagram that d<a; otherwise we can repeat the same analysis and construction starting with the reflection of AB in AC).

Then $BE = AB - AE = AB - AD = a - d$, and CE is congruent to CD as its reflection image. Hence we know all the sides in $\triangle BCE$: $BE = a - d, BC = b, CE = CD = c$, and we can construct this triangle by SSS.

CONSTRUCTION. Construct $\triangle BCE$ with the sides $BE = a - d, BC = b, CE = c$. Extend BE beyond E by a segment $EA = a - d$. Draw AC. Reflect $\triangle BCE$ in AC. Let point D be the reflection image of B. $ABCD$ is a sought for quadrilateral.

PROOF. It follows immediately from the construction.

INVESTIGATION. $\triangle BCE$ is defined uniquely by SSS. Point A is defined uniquely by AXIOM 3. Point D is defined uniquely as the reflection image of point E. Hence, with the given choice of the order of sides and the suggestion of which angle is bisected by the diagonal, the solution is unique.

7. Review.

28. Let segment a be a lateral side and h the altitude to that side in an isosceles triangle. We have to construct the triangle.

a h

Solution 1 (A quick one).

ANALYSIS.

A sought for triangle can be viewed as half of a rhombus with the lateral side *a* and the altitude *h*, as shown in the figure below.

If $h < a$, there are two possible solutions: ΔACD with the vertical angle (angle at the vertex) acute or ΔDBC with the vertical angle obtuse (in this case the altitude falls onto the extension of a lateral side, which is shown as a dotted line in the diagram.

If $h = a$, the rhombus will be a square, and the sought for triangle will be right.

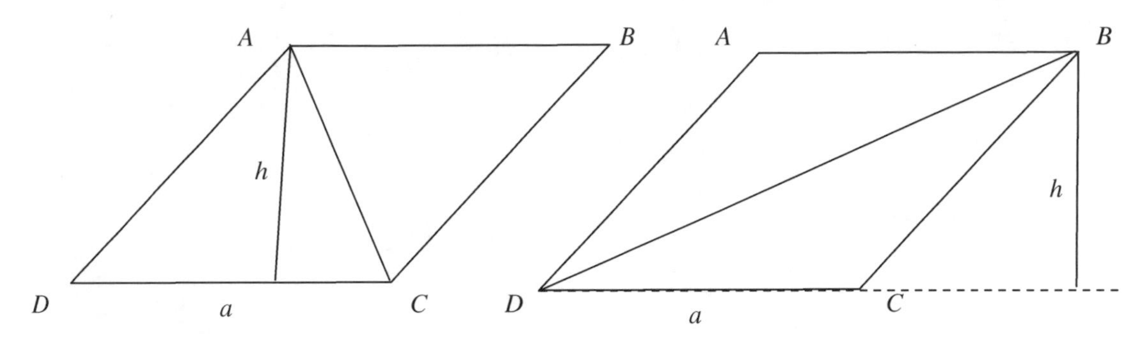

CONSTRUCTION.

Draw two parallel lines *l* and *m* standing at the distance *h* from each other (this can be done, for instance, by drawing two lines perpendicular to the given segment *h* and passing through its endpoints).

Then, from an arbitrary point *A* on one of these lines (line *l* in the diagram below), describe a circle of radius *a*. It will cut *l* at two points, labeled *E* and *B*.

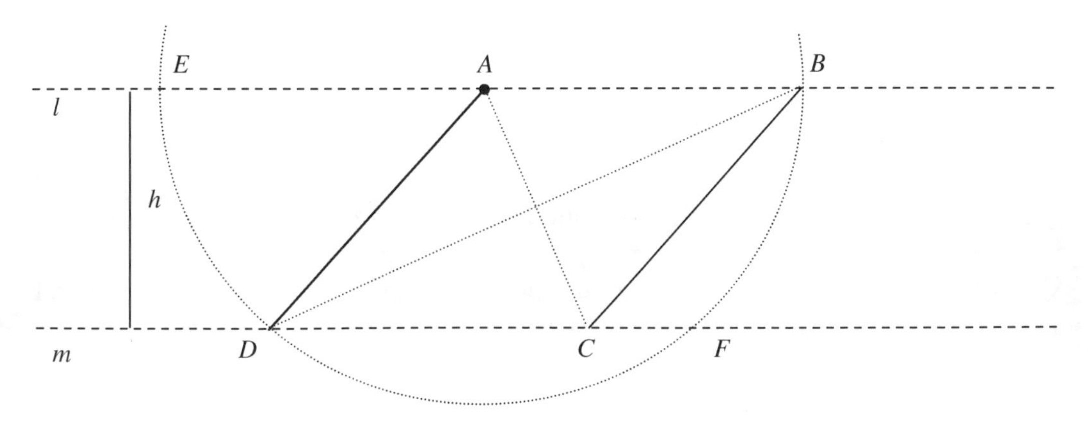

If $h < a$, it will cut *m* at two points; let us denote them *D* and *F*. Through *B*, draw a line parallel to *AD*. It will intersect *m* (how do we know it will?) at some point *C*. By construction, *ABCD* is a parallelogram with two neighbouring sides *AD* and *AB* congruent as radii of the same circle. Since opposite sides in a parallelogram are congruent, it follows from $AD=AB$ that $BC=AD=AB=CD$; hence $ABCD$ is a rhombus. Then either of the triangles ADC or DAB is a solution: each of them is isosceles ($AD=CD$ or $AD=AB$) with an altitude h to either of congruent sides.

If $h=a$, the circle of radius a described from A will touch line at a single point D, and the rhombus constructed according to the above procedure will be a square with the side $h=a$. In this case triangles ADC and DAB will be right congruent triangles.

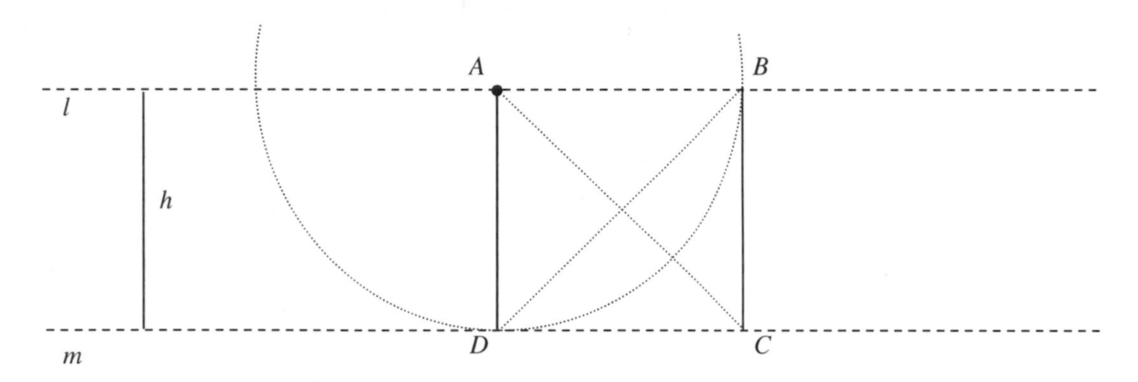

SYNTHESIS – it is already contained in the construction.

INVESTIGATION.

If $h=a$, the solution is unique, since every other isosceles right triangle with the same leg a will be congruent to the above ΔDAB by *SAS* (or by two legs).

If $h<a$, the required triangle will not be right: the vertical angle will be acute or obtuse. Suppose some triangle satisfies the required conditions; let us complete this triangle into a rhombus by attaching to it base-to-base a congruent triangle. Then we shall obtain a rhombus $MNQP$ in which the distance between the opposite sides is h. Let us place this rhombus as well as $ABCD$ that we have constructed above in such a way that their opposite sides will lie on two parallel lines l and m standing at the distance h from each other.

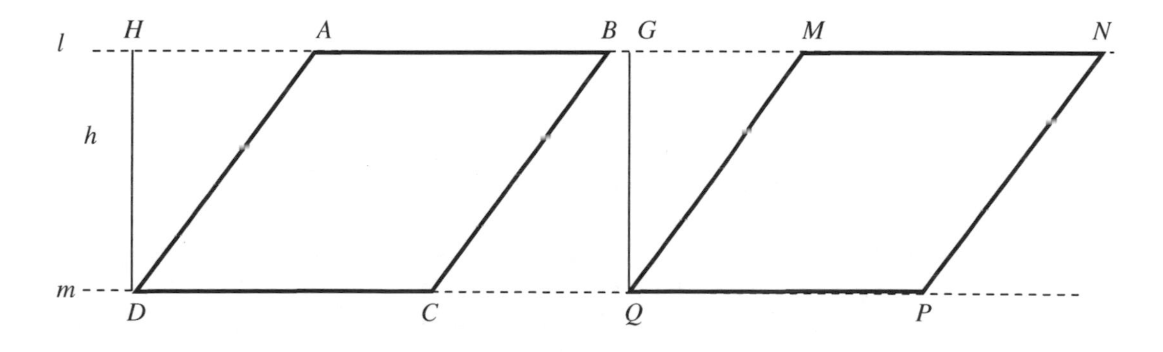

If H and G are the feet of the perpendiculars dropped upon l from D and Q respectively, then right triangles DHA and QGM are congruent by a leg and hypotenuse.. Let us subject the plane to an isometry that will impose ΔQGM onto ΔDHA. Then QM will coincide with AD. Since $MN=a=AB$ and both segments lie on the same line, MN will coincide with AB; N will fall onto B. Then NP will coincide with BC since there is only one line through B parallel to AD, and the segments are congruent. Thus $MNPQ$ will coincide completely with $ABCD$, which means they are congruent.

Hence, the triangle that is a solution to the problem will be congruent to DAC or DBC; thus the problem will have two solutions when $h < a$.

If $h > a$ the problem will have no solutions since the perpendicular to a line cannot be greater than an oblique drawn from the same point to the same line.

Solution 2 (A straightforward one).

ANALYSIS.

Let ABC in the figure (i) below be a sought for triangle with the lateral sides $AB = BC = a$ and an altitude $AD = h$ to one of the lateral sides.

We can construct a right triangle ABC with a leg h and hypotenuse a. Such a triangle (shown in grey in the diagram can be easily completed into a sought for triangle.

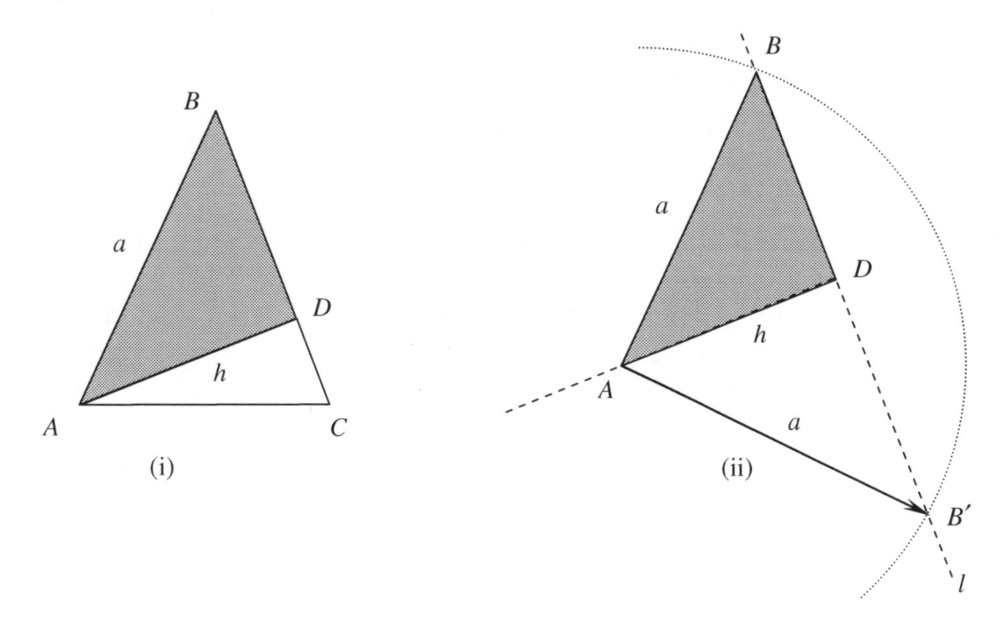

CONSTRUCTION.

Let us first consider the case $h < a$.

From an arbitrary point D on some line l, erect a perpendicular and lay of on it $DA=h$ (see diagram (ii)). From A as a centre, describe an arc of radius a; it will cut line l at two points: B and B'. From B, describe a circle of radius a; it will cut l at two points : C and E (see the diagram below).

By construction, each of the triangles ABC and ABE is a solution, since each of them is an isosceles triangle with a lateral side a and an altitude to a lateral side h.

If $h = a$, construct a right triangle with each leg congruent to a (trivial).

SYNTHESIS. It is included in the construction.

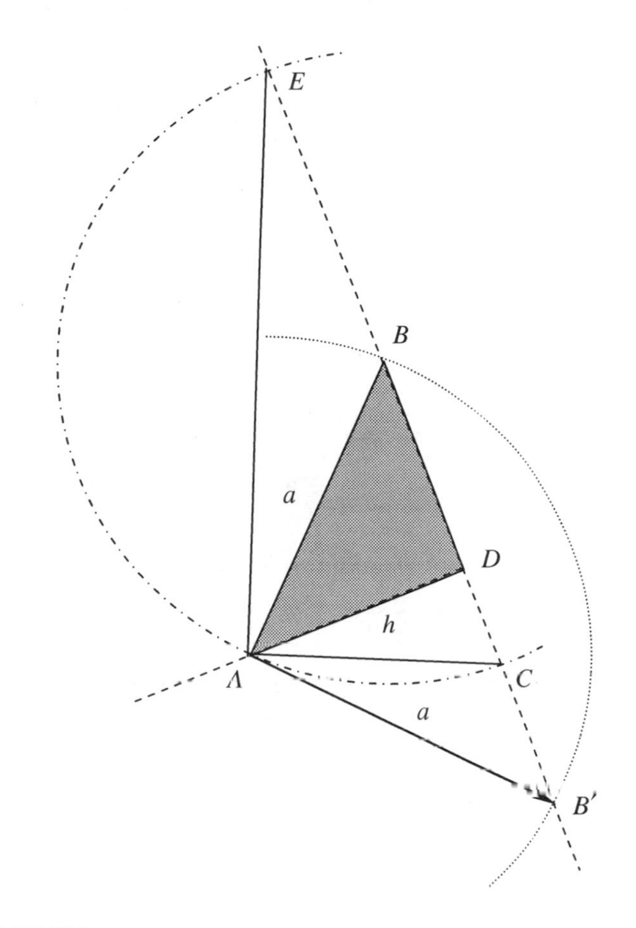

INVESTIGATION.

If there is an isosceles triangle with the same lateral side a and an altitude h to that side, and $h < a$, perform an isometry that will make the right triangle formed by its lateral side and altitude to coincide with ΔABD in the above diagram. Then, by Axiom 3, the other lateral side of the triangle will coincide with *BC* or with *BE*, so the base will fall onto *AC* or *AE* respectively.

Thus the triangle will coincide with one of the two constructed triangles. Then there are exactly two solutions if $h < a$.

There is exactly one solution if $h = a$, and no solutions if $h > a$ (both proofs are trivial).

Additional question: Is solution 2 valid in Neutral Geometry?

32. Let *b* be the given base, α – an angle with the vertex at an endpoint of the base, and *h* – the altitude dropped onto the base.

The solution is not dissimilar to solutions of problem 28. We shall briefly describe it.

CONSTRUCTION.

Let us draw two parallel lines *l* and *m* standing at the distance *h* from each other (this can be done, for instance, by drawing two lines perpendicular to the given segment *h* and passing through its endpoints).

From an arbitrary point *A* on one of this lines (*m*) draw a ray *k* that makes angle α with the line and lies in the same half-plane as the other line (*l*). This ray will intersect the other line (why?) at some point *B*. From *A*, lay off a segment $AC=b$. Draw *BC*.

Triangle *ABC* is a sought for, and the solutions is unique (the proof is trivial: show that any other triangle satisfying the given conditions is congruent to the one we have constructed).

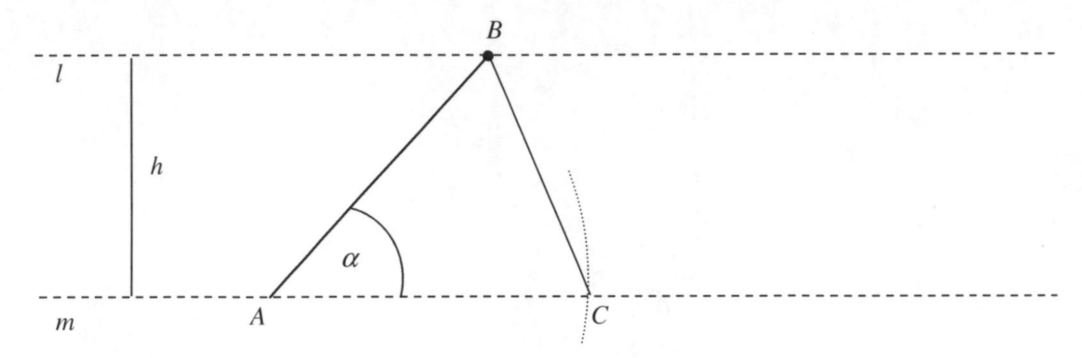

Sections 8.1-2

10. Let *AB* be a diameter in a circle centered at *O*, and *AC* and *BD* two parallel chords drawn through the endpoints *A* and *B* of this diameter. Let us show that *AC=BD*.

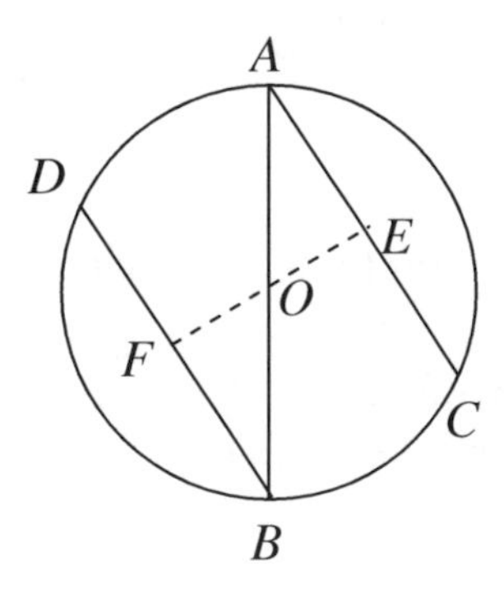

From *O*, drop a perpendicular *OE* onto *AC*; then according to Corollary 1 of Theorem 6.2.2, the line containing *OE* will be perpendicular to *DB* as well. Let *F* be the foot of this perpendicular to *DB*.

Then triangles *OFB* and *OEA* are congruent by ASA condition: $\angle A = \angle B$ as alternate angles formed by the parallel chords and their transversal *AB*, each of the sides *OA* and *OB* is congruent to the radius, and thus they are congruent, and the angles *FOB* and *EOA* are congruent as verticals.(One could also notice that these two triangles are congruent as two right triangles by an acute angle and the hypotenuse.)

It follows from the congruence of these triangles that *OE*=*O*F, and therefore the chords *AC* and *DB* are congruent as the chords located at the same distance from the centre of the circle (Th.8.2.2).

13. If the given point is the centre, any chord through it will be a diameter and hence it will be bisected by the point. In this case the problem has infinitely many solutions.

Otherwise (if the given point is not the centre), draw a line through that point and the centre (by AXIOM 1, there exists exactly one such line), and draw a line perpendicular to the first line through the given point (also, there will be exactly one such a line- why?). The part of this line that lies inside the circle, will be by Th.8.1.2, the sought for chord. In this case the problem has a unique solution, since a chord is bisected by a diameter iff it is perpendicular to the diameter, and there exists only one line through the given point that is perpendicular to the diameter passing through the given point.

22. The trapezoid must be isosceles, as it follows directly from Theorem 8.1.4.

Section 8.3

11. Let us consider two tangents drawn from some point A to a circle centered at some point O, as shown in the figure below. Let B and C be the points at which the tangents touch the circle.

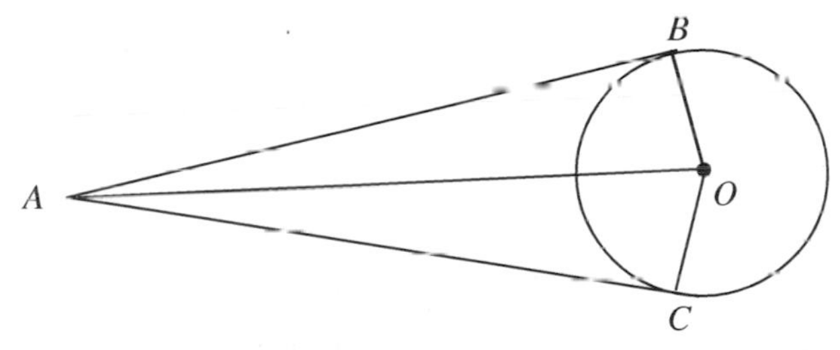

By Theorem 8.3.2, the radii OB and OC drawn to the points of contact B and C are perpendicular to the respective tangents; hence triangles AOB and AOC are right. These triangles have a common hypotenuse AO, and $OC=OB$ as the radii of the same circle. Therefore these triangles are congruent, and $AC=AB$ as corresponding legs.

Thus, the segments of two tangents drawn to a circle from a point, are congruent.

14. Let $P(\Delta MKL)$ denote the perimeter of ΔMKL. According to the result proved in problem 11, $KA=KC$ and $LB=LC$ as the segments of the tangents drawn from the same point; hence

$$P(\Delta MKL) = ML + MK + LK = ML + MK + LC + KC = ML + MK + LB + KA =$$
$$= ML + LB + MK + KA = MA + MB.$$

Therefore, the perimeter is equal to the sum of the segments of the tangents drawn from M, and it does not depend upon the location of C, Q.E.D.

Section 8.5

15. Let ABC be a triangle inscribed in a circle, and the angles of this triangle are: $\angle A=\alpha$; $\angle B=\beta$; $\angle C=\gamma$. Then one of the following three situations will take place: One of the angles is obtuse; (b) all three angles are acute; (c) one of the angles is right.

(a) One of the angles, e.g. α, is obtuse. In this case, illustrated below in the diagram (a), according to Theorems 8.5.1 and Lemma 8.5.3, $\angle PCA=\angle PAC=\beta$, since each of these angles is formed by a tangent and a chord AC that subtends an inscribed $\angle ABC=\beta$. Similarly, $\angle NBA=\angle NAB=\gamma$.

$\angle NBC+\angle PCB=\gamma+\beta+\gamma+\beta=2(\beta+\gamma)$. Since in the case under consideration $\angle BAC=\alpha$ is obtuse, then $\beta+\gamma=2d-\alpha<d$; $\Rightarrow$ $2(\beta+\gamma)<2d$; $\Rightarrow$ $\angle NBC+\angle NCB<2d$. Hence the tangents touching the circle at B and C will intersect, according to the Euclidean parallel postulate, at some point M, lying "behind" points N and P, i.e. on the extensions of BN and CP beyond N and P respectively. Then the circle will be *escribed* (see section 8.8) to triangle MNP formed by the tangents and A, B, and C.

Then $\angle NPM=2\beta, \angle MNP=2\gamma$ as exterior angles for the triangles CPA and ANP respectively, and $\angle M=2d-2\beta-2\gamma=2d-\beta-\gamma-(\beta+\gamma)=\alpha-(2d-\alpha)=2\alpha-2d$. (The same result could be obtained by applying Theorem 8.5.4).

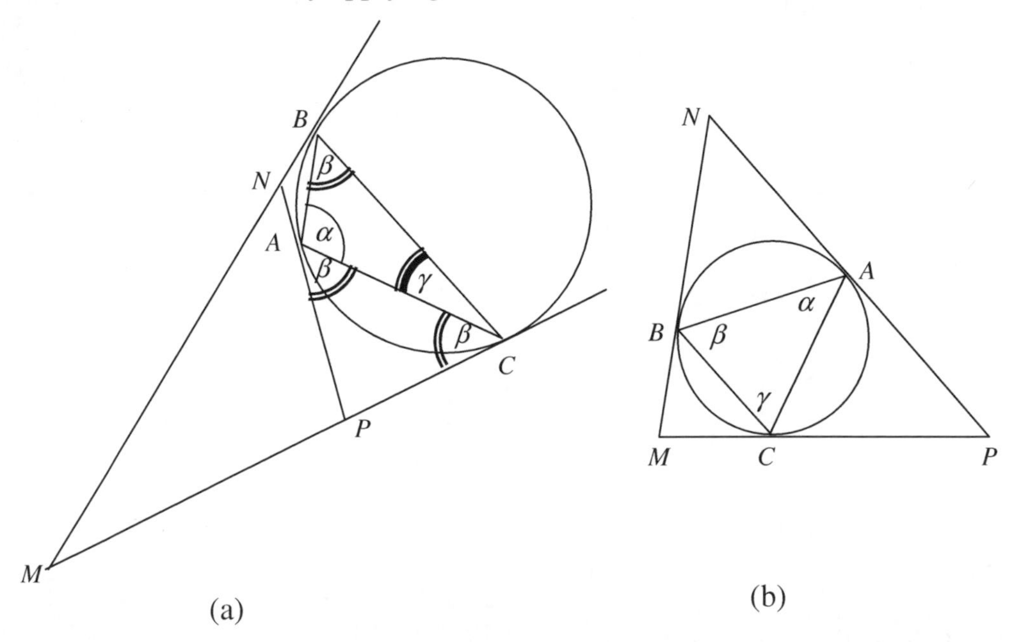

(b) In this case, illustrated in the diagram (b) above, the circle will turn out to be *inscribed* in the triangle MNP formed by the tangents.

By Theorem 8.5.3, $\angle MBC=\angle MCB=\alpha$, whence $\angle M=2d-2\alpha$. (We could use Theorem 8.5.4 to obtain the same result).

Similarly, $\angle P=2d-2\beta$; $\angle N=2d-2\gamma$.

(c) If $\angle A$ is right, BC will be a diameter, and the tangents drawn through B and C will not intersect as perpendiculars to the same diameter. Then a triangle will not be formed.

Section 8.6

12.
ANALYSIS. The locus of all the centres of the circles inscribed in an angle is the bisector of the angle. Also, if a circle touches a line at some point, it must lie on the perpendicular to the line at that point.
CONSTRUCTION. Draw the bisector of the given angle and erect a perpendicular from a given point on a side to that side. The point of their intersection is the centre of the sought for circle. Describe the circle by the radius congruent to the distance from that centre to the given point on a side.
PROOF is already contained in ANALYSIS.
INVESTIGATION. Since an angle is tacitly suggested to be less than two right angles, the sum of the interior angles formed by the side of the angle intersecting the perpendicular erected to it in the interior and the bisector of the angle will be less than two right angles; hence, by the Euclidean parallel postulate, they will intersect. Two lines may intersect only at one point, so the solution is unique.

15. Given a point in the exterior of a circle, draw a secant whose interior part equals to a given segment.
Let A be a point in the interior of a circle with the centre at some point O, as shown in the diagram, and segment a be the given segment.

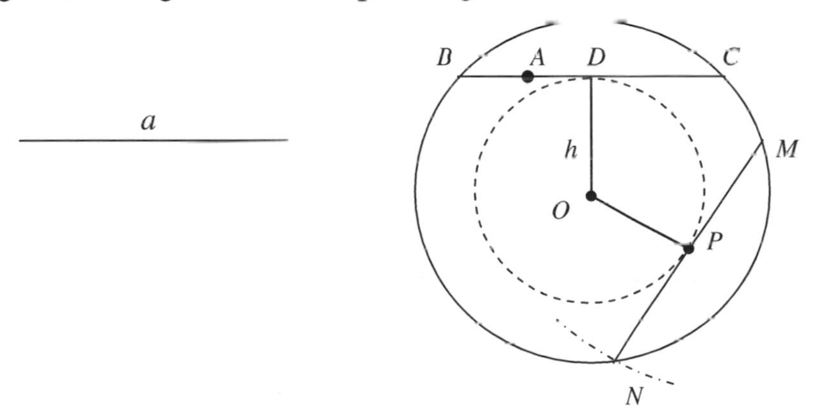

ANALYSIS. Let BC be a sought for secant, which means it passes through A and $BC=a$.

According to Theorem 8.2.1, every secant congruent to a (it would be more precise to say: "every secant from which the circle cuts off a segment congruent to a") will stand at the same distance $OD=h$ from the centre of the circle. Therefore, the circle of radius h with the centre at O will be tangent to all such secants.

Then the plan is the following: first we shall determine h and describe a circle of radius h from O. this circle is the locus of points standing at the distance h from O, hence all the secants congruent to a will be tangent to this circle. Then we shall draw tangents to that circle from point A.
CONSTRUCTION. From an arbitrary point M on the given circle, describe an arc of radius a. If a does not exceed the diameter of the circle, the arc will cut (or touch) the

circle at some point *N*. From *O*, drop a perpendicular *OP* onto *MN* (or find the midpoint *P* of *MN*).

From *O* as a centre, describe a circle of radius congruent to *OP*. Through the given point A, draw a tangent to that circle. By Th.8.2.1, the segment of this tangent line enclosed within the given circle, will be congruent to *MN=a*.

PROOF. It is already contained in the construction.

INVESTIGATION. If *a* is less than the diameter, there are two solutions, since there are two tangents to a given circle from a given point.

If *a* is congruent to the diameter, there is a unique solution, namely the diameter drawn through *A*.

If *a* is greater than the diameter, there is no solutions, since diameter is the greatest chord in a circle.

20.

ANALYSIS. The line passing through the centre of the given circle and the point of tangency is the locus of all the centres of the circles touching the given circle at the given point; so the centre of the sought for circle must lie on that line.

On the other hand, the centre of the sought for circle must be equidistant from the given point of tangency and the other given point, through which it passes. Therefore it must also lie on the perpendicular bisector of the segment joining the given points.

CONSTRUCTION. Let the *B* be the given point on the given circle centered at *O*, and *A* be the given point that must lie on the circle that is to be drawn. Draw line *OB* and the perpendicular bisector of *AB*. The point *C* of their intersection is the centre of the required circle.

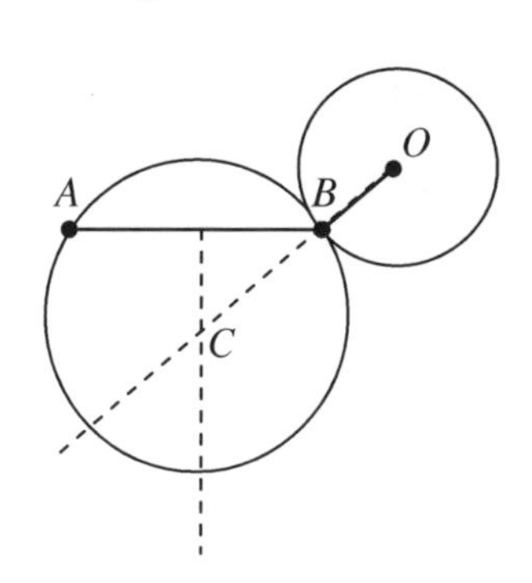

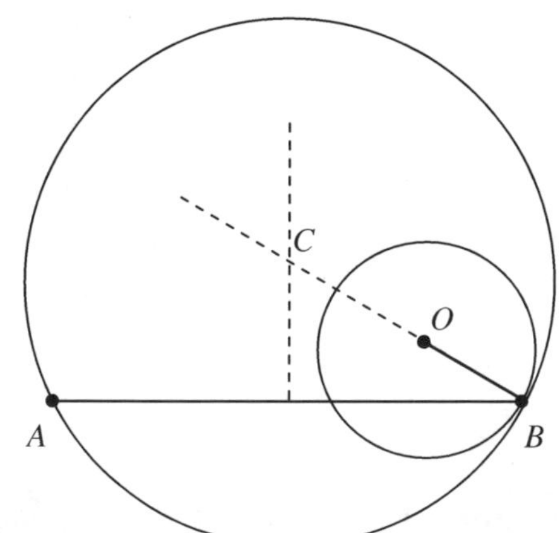

INVESTIGATION. The two loci will not intersect if *OB* is perpendicular to *AB*, in which case there is no solutions; otherwise (if they form an acute or obtuse angle) there will be a unique solution, since the two loci are straight lines and they will intersect at exactly one point, since they are not parallel. If the points *O*, *A*, and *B* are collinear, there will be a unique solution as well: the centre of the sought for circle will be located at the midpoint of *AB* (which will be a diameter of the sought for circle).

23. Describe a circle of a given radius such that it touches a given line and a given circle.

Let r be the given radius of a sought for circle, which must touch the given line l and the circle of radius R with the centre at O , shown in the figure below.

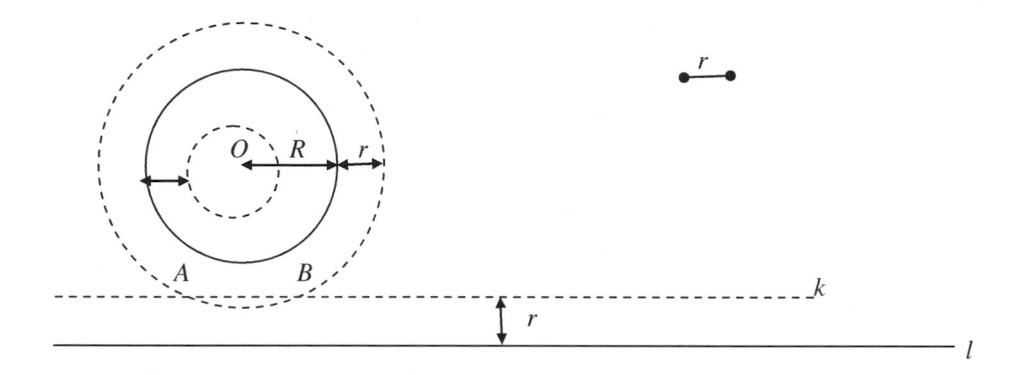

ANALYSIS. Since a sought for circle must have line l as a tangent, its centre will lie at the distance r from line l. The locus of all the points having this property is a straight line k parallel l passing at the distance r from l.

Also, the centre of a required circle will lie at the distance r from the given circle. The locus of all such points consists of two circles: one of radius $R+r$, and the other of radius $R - r$ (or $r - R$ if r is greater than R, as shown in the diagram below, where the circles touch internally with the given circle enclosed in the interior of the sought for circle) concentric with the given circle.

The centre of a sought for circle will be a point of intersection of the two aforementioned loci, which are shown in dotted lines in the diagram.

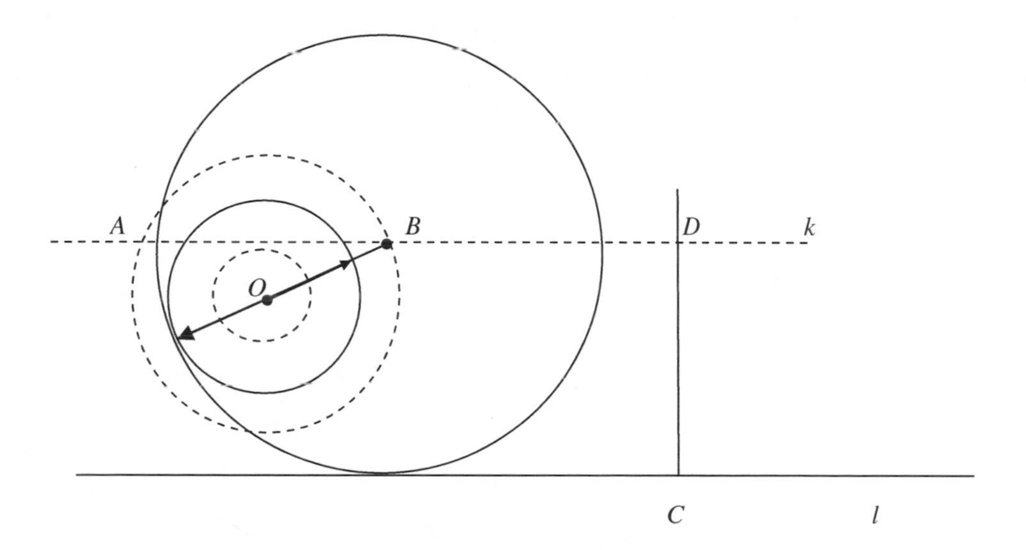

CONSTRUCTION. Draw a line k standing at the distance r from l (from an arbitrary point C on l, erect a perpendicular p to l, lay off on this perpendicular $CD=r$, and draw a line k through D parallel to l).

From O as a centre, describe a circle of radius $R+r$.

If this circle and the line k have a common point (A or B in the diagram below), this point is the centre of a sought for circle. From this point describe a circle of radius r.

PROOF. It is trivial (can be omitted, since it is already contained in the analysis).

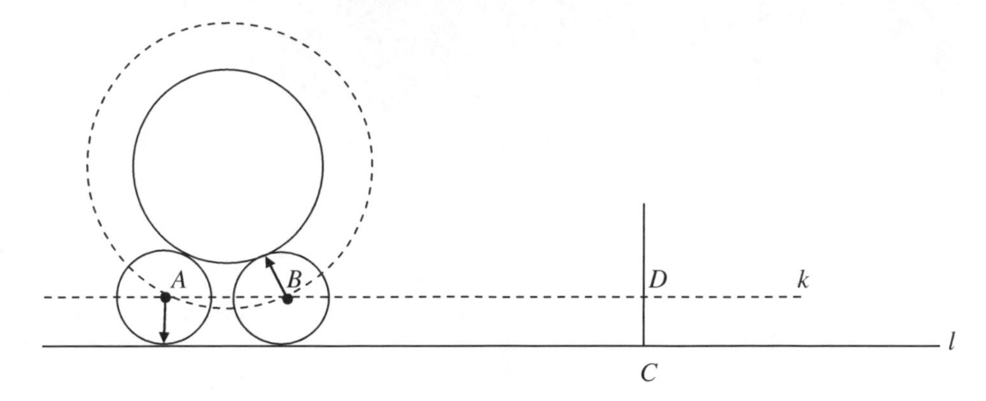

INVESTIGATION. The problem may have from zero up to four solutions: there is no solutions if the dotted line k does not have common points with any of the dotted circles, and four solutions if the dotted line k intersects both dotted circles.

The latter situation is illustrated in the diagram below, where two of the four solutions (the circles with their centers at E and F) are drawn in bold lines, and the other two solutions, with their centers at A and B , are not shown.

The number of solutions depends upon the relations between the given radii and distances between the given circle and the given line.

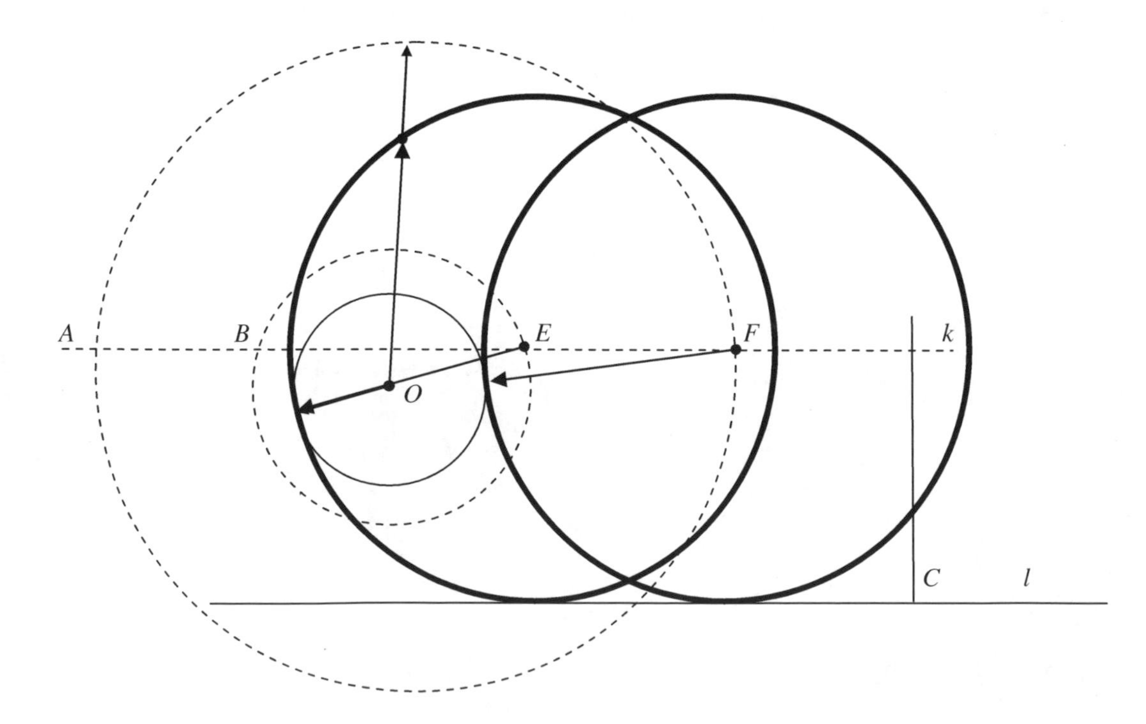

In particular, if r is less than R, there will be:

2 to 4 solutions if the distance between the centre O of the given circle and the given line l is less than $R+2r$ (in this case line k will intersect at least one of the dotted line circles at two points, A and B) ;

one solution if this distance is exactly $R+2r$ (in this case line k will be tangent to the dotted line circle) ;

no solutions if this distance is greater than $R+2r$.

24. Let $\angle A$, formed by the rays l and m emanating from point A be the given angle, and the segments a and b be the given segments congruent to the chords cut off from l and m by the sought for circle. Let r be the given radius of the sought for circle.
ANALYSIS. If the centre of the circle is located at some point O, as shown in the diagram below, it is clear that O must stand at certain distances from the sides of the angle. The distances, denoted h and H, are the distances from the centre O of the required circle to the chords $BC = a$ and $DE = b$ that are cut off on the sides.

The locus of all the points standing at the distance h from the side l is a line parallel to l such that the perpendicular dropped from any point of this line upon l is congruent to h.

Therefore, the centre O of the sought for circle must be located on such a line, shown as a dotted line parallel to l on the right diagram below. Similarly, O is located on a line parallel to m and standing at the distance H from it (another dotted line). Hence, O can be found as a point of intersection of the two loci.

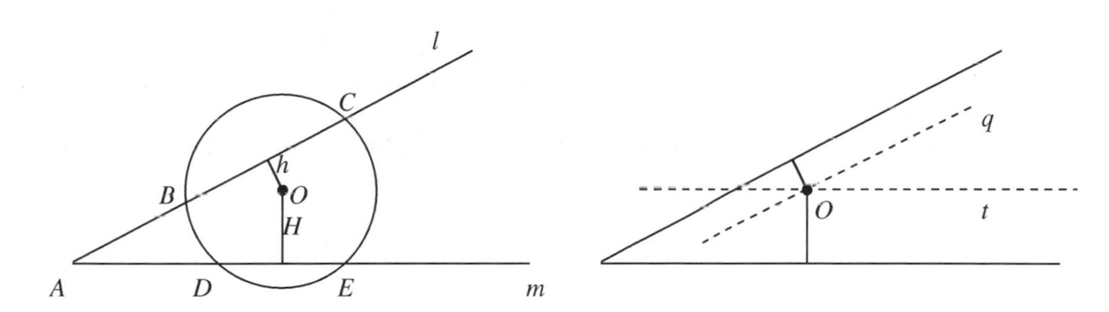

CONSTRUCTION. First of all we have to find the distances h and H from the centre of the circle to the chords congruent to a and b respectively. For definiteness, we suggest that a is to be cut off on l and b – on m, and $a \geq b$. (Then, by Th.8.2.1, $h \leq H$).

Let us draw a circle of the given radius r with its center at some point Q. (See the left diagram below). From an arbitrary point M on this circle, describe an arc of radius a. (We suggest that $a \leq 2r$, otherwise it cannot be a chord in a circle of radius r.) The arc will cut the circle at some point N. Draw MN. From the centre of the circle, drop a perpendicular upon MN. This perpendicular QP is the distance from the centre of the circle to the chord $MN = a$, i.e. in our notations $h=QP$. (Let us notice that the locus of all the points standing at the distance h from the centre will be a concentric circle of radius $h=QP$ (see the right diagram below), and any chord tangent to this circle will be congruent to a, according to Th.8.2.1).

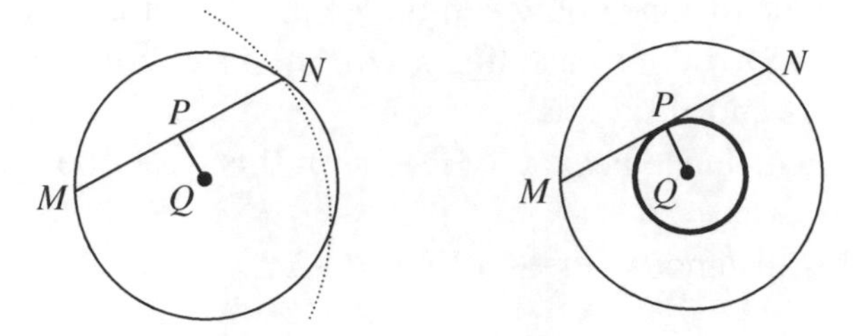

Similarly, we can construct the distance H from the centre O to the chord congruent b.

Draw a line q parallel to l and standing at the distance h from it. (This can be done by erecting a perpendicular from any point on l, laying off on this perpendicular a segment congruent to h, and drawing a perpendicular to the first perpendicular through the second endpoint of the segment congruent to h). Let us notice that we can draw two such lines: p and q, lying on the opposite sides of l, as shown in the left diagram below. Similarly, we can draw a line standing at the distance H from m (it is also shown as a dotted line in the diagram).

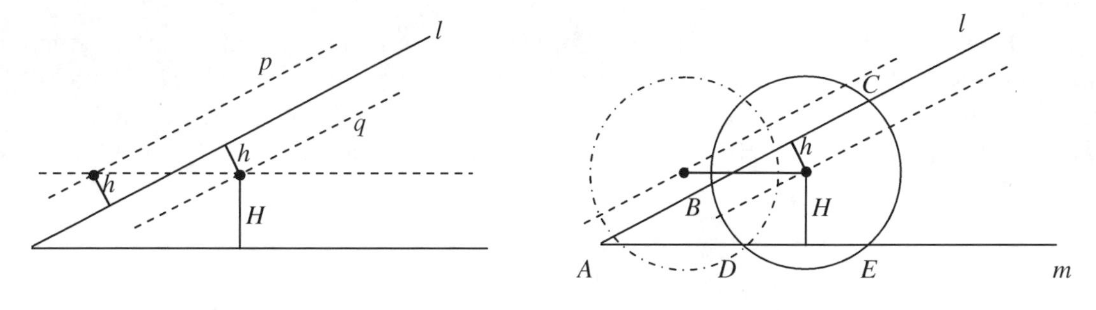

The point of intersection of the latter line with p or q is the centre of the sought for circle.

PROOF is trivial: it repeats the substantiation of ANALYSIS and CONSTRUCTION.

INVESTIGATION. As we have already mentioned, each of the given chords must be less than the diameter $2r$ of the circle. Under this condition, the problem may have up to two solutions, depending on the relations between the chords, radius, and on the given angle.

The right diagram above illustrates the case of two solutions: each of the lines, p and q, intersects with the line standing at the distance H from the other side of the given angle, and each point of intersection satisfies the requirements for the centre of the circle. Still for some configurations (try to imagine one with a greater given angle), only one solution exists. It is easy to prove that at least one solution exists if the given angle is acute.

If the given angle is obtuse, the problem may have one solution or no solutions at all, as illustrated in the diagrams below.

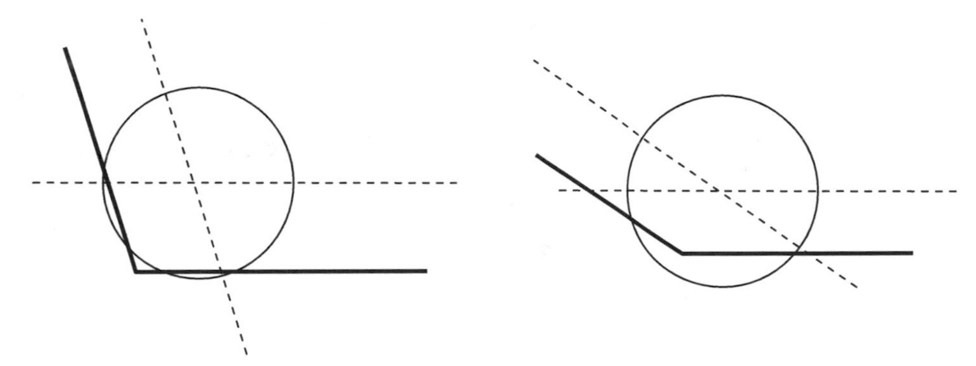

One can try to find exact conditions that will determine the number of solutions (use the concentric circles that are the loci of the endpoints of the distances from the centre to the given chords, mentioned above when h was being constructed).

31. Through a point of intersection of two given circles, draw a secant with the sum of its segments lying in the interiors of the circles equal to a given segment.

Let two circles with the centers at O and Q and radii r and R respectively be given, and these circles intersect, P being one of the points of intersection, and we have to draw through P a common secant of the circles, such that the sum of the segments of the secant lying inside the circles is equal to the given segment l.

ANALYSIS Let CD in the figure below is a sought for secant, i.e. $CP + PD = l$.

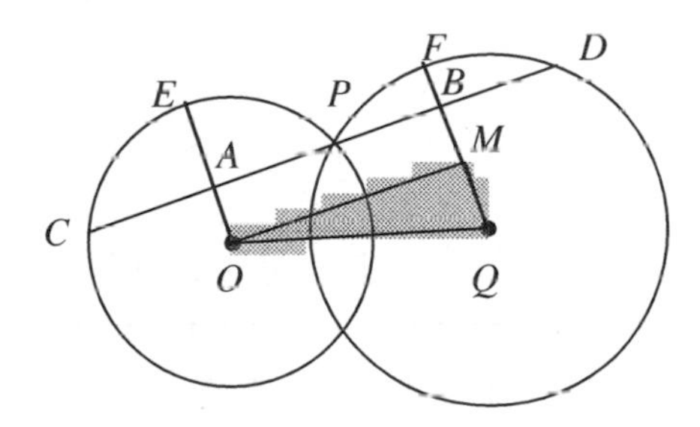

Let OE and QF be the radii perpendicular to CD; then since they are perpendicular to the chords CP and PD, they cut these chords at the midpoints A and B (Theorem 8.1.3).

Hence $AB = AP + PB = \frac{1}{2}CP + \frac{1}{2}PD = \frac{1}{2}(CP + PD) = \frac{1}{2}l$.

Let us drop a perpendicular OM from O onto QF. Then $OM = AB$ as two perpendiculars between two parallel lines OE and QF (the latter are parallel since they are both perpendicular to CD). Therefore, $OM = AB = \frac{1}{2}l$, and we can construct $\triangle OMQ$ by its hypotenuse OQ and a leg $OM = \frac{1}{2}l$.

After this triangle has been constructed, we shall know the location of point M. Then we shall draw QM, and through P we shall draw a line perpendicular to OM. This line will be a sought for secant.

CONSTRUCTION. First we shall construct a right triangle congruent to $\triangle OMQ$:

On a ray *k* emanating from some point *S*, we lay off a segment $ST = \frac{1}{2}l$. From *T*, we erect a perpendicular *p* to *k*. From *S*, we describe a circle of radius congruent to *OQ*. If $OQ > \frac{1}{2}l$, the circle will intersect perpendicular *p* at some point *U*; otherwise (if $OQ > \frac{1}{2}l$) the right triangle cannot be constructed (the hypotenuse must be greater than a leg), and the problem has no solutions.

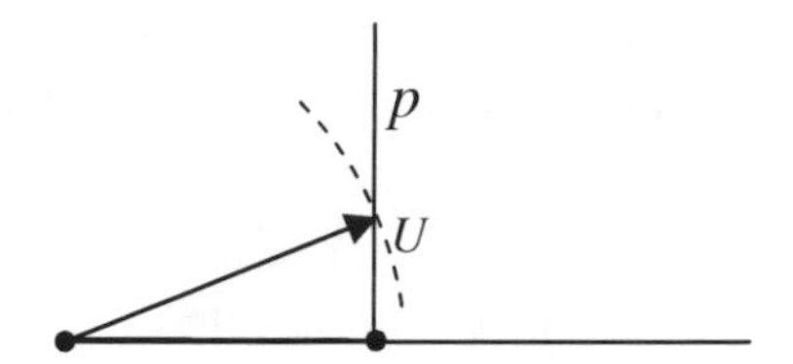

From point *O*, draw an arc of radius *ST*. From point *Q*, draw an arc of radius *TU*. The arcs will intersect (why?) at some point *M*.

Draw line *OM*. Through *P*, draw a line perpendicular to *OM*. This line is a sought for secant.

PROOF. It is already proved in ANALYSIS that $CD = l$.

INVESTIGATION. Suppose there is another secant through *P* that has the same property. Then let us draw radii perpendicular to this secant and drop a perpendicular from each centre of a circle onto the radius of another circle. Thus we shall form two right triangles, one of which will coincide with $\triangle OMQ$ from the figure provided in ANALYSIS, while the other, with the "short" leg *ON* emanating from *O*, will be located as shown in the figure below. By drawing through *P* a secant parallel to the leg *NQ* of the shaded right triangle, as shown in the diagram below, we obtain the second solution of the problem.

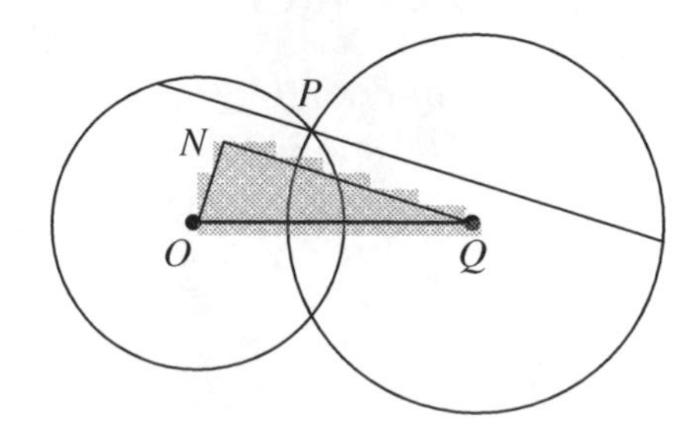

For some relations between the segments OQ and l the problem may have one solution or no solutions. The latter is the case when $OQ < \frac{1}{2}l$, since we will not be able to construct $\triangle OMQ$; also there will be no solutions if point P falls into the interior of both triangles OMQ and ONQ, as shown in the diagrams below. If P falls in the interior of only one of these triangles (ONQ if $R>r$), then the problem will have exactly one solution.

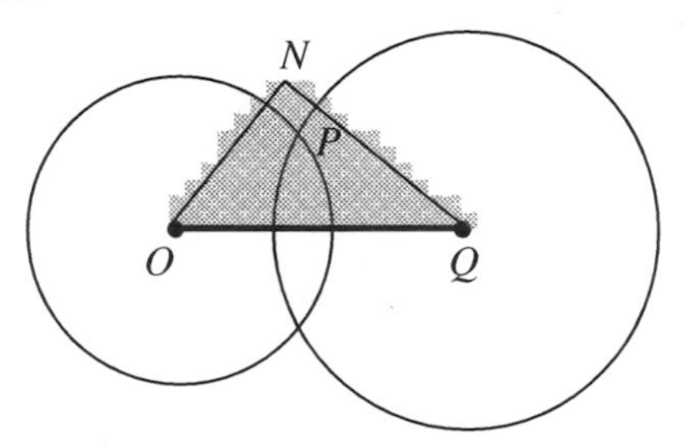

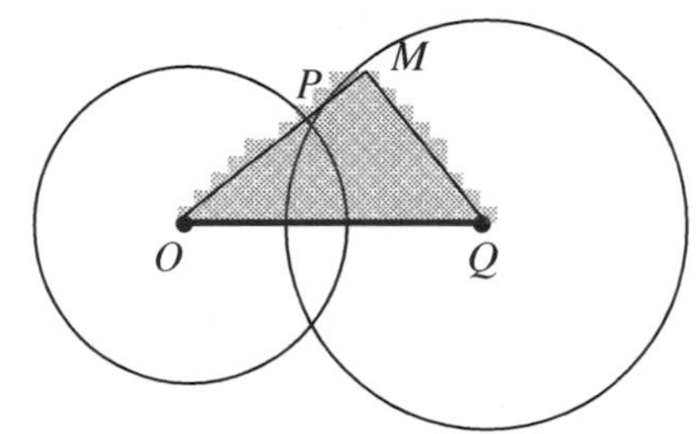

Section 8.7

7. Let the bisectors of the exterior angles CBE and DCB intersect at some point O. The bisector of an angle is the locus of points equidistant from the sides of the angle. Since BO is the bisector of $\angle CBE$, point O is equidistant from the lines BE and BC, and since CO is the bisector of $\angle DCB$ point O is equidistant from BC and CD.

Hence O is equidistant from CD and BE, which lie on the rays AD and AC. Thus, O must lie on the locus of points equidistant from the sides AC and AB of angle BAC, therefore it lies on the bisector of angle BAC, Q.E.D.

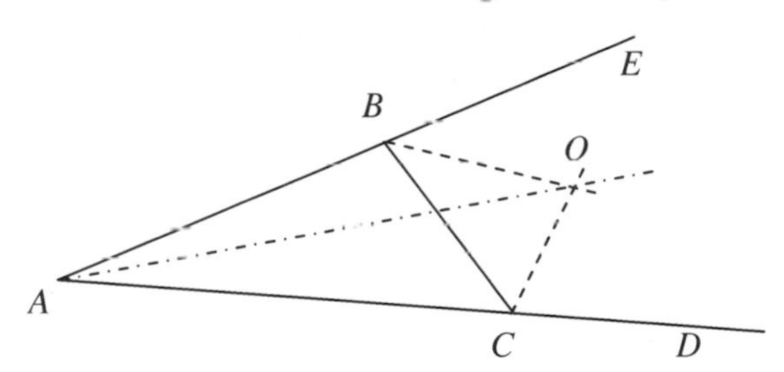

Section 8.8

10. Hint: Construct by SSS a triangle formed by the given base and the segments congruent to two thirds of the given medians (this triangle is painted grey in the diagram below). Then extend the lateral sides of the obtained triangle to find the midpoints of the lateral sides of the sought for triangle.

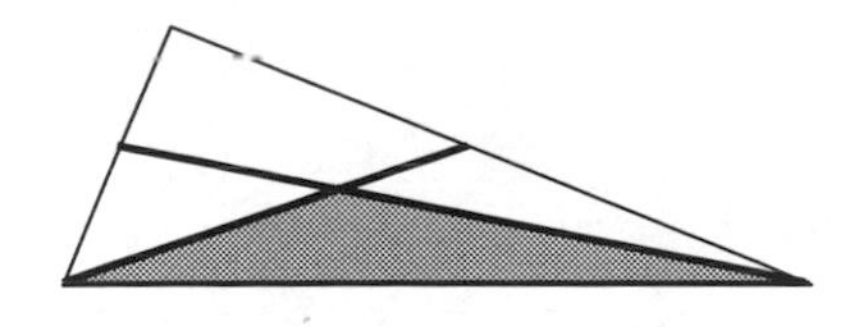

17. Each vertex stands from the centroid at the distance equal to two thirds of the median emanating from it, hence the vertex from which the least median emanates will be the closest to the centroid.

Let us show that the least median emanates from the vertex opposed the greatest side of a triangle.

Let in triangle *ABC* in the figure below *AB*>*BC*, and *AD*, *BE*, and *CF* are the medians; according to Theorem 8.8.2, they are concurrent at point *O*, the centroid.

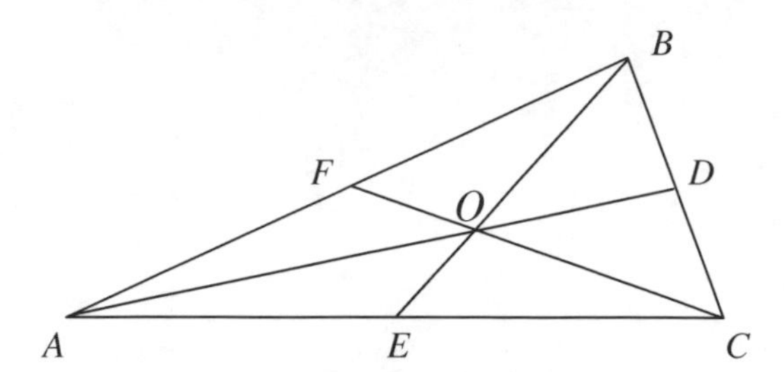

In triangles *ABE* and *CBE*, *EA*=*EC* (since *BE* is a median), *EB* is a common side, and *AB*>*BC*; hence according to Theorem 4.7.3ii, $\angle BEA > \angle CEA$.

Now let us consider the triangles *AOE* and *COE*. In this triangles *EA*=*EC*, *OE* is common, and as we have shown above, $\angle BEA > \angle CEA$; hence by Theorem 4.7.3i, *AO*>*CO*. Therefore the centroid will be closer to the vertex opposed to the greater of two sides (in our case *C* is closer than *A* to the centroid).

Since by Theorem 8.8.2 $AO = \frac{2}{3}AD = \frac{2}{3}m_a$, and $CO = \frac{2}{3}CD = \frac{2}{3}m_c$, it follows that *AD*>*CF*, i.e. the median drawn to the lesser side is greater than the median drawn to a greater of two sides.

Section 9.1

9. Find the *GCM* (27*cm*, 111*cm*).

$111cm = 4\cdot(27cm) + 3cm;$

$27cm = 9\cdot(3cm); \quad \Rightarrow GCM(A,B) = 3cm$

10. According to the Euclidean algorithm, the greatest common measure of the given segments *A*=246*U* and *B*=16*U*, is determined by the following sequence of divisions with remainders:

$246U = 15\cdot 16U + 6U$

$16U = 2\cdot 6U + 4U$

$6U = 1\cdot 4U + 2U$

$4U = 2\cdot 2U.$

Therefore, GCM (*A*,*B*)=2*U*.

16. In an isosceles triangle with the measure of the vertical angle 36^0, each base angle (an angle whose side coincides with the base) will measure $\frac{1}{2}\left(180^0 - 36^0\right) = 72^0$.

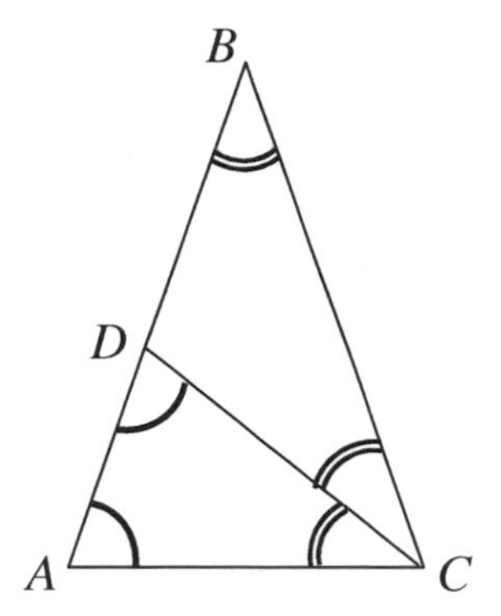

Let in ΔABC with $\angle ABC = 36^0$, $AB = BC = l; AC = b$.

It should be noticed that AC is the least side of the triangle (why?). Then, according to the Euclidean algorithm,

$l = nb + r; \quad r < b.$

We shall start with finding such a number n. Draw the bisector CD of $\angle ACB$. Then, since $\angle ACD = \frac{1}{2}\angle ACB = 36^0$, $\angle ADC = 180^0 - 72^0 - 36^0 = 36^0 = \angle DAC$, hence $\triangle ADC$ is isosceles, with $CD = AC = b$.

Since $\angle DCA = 36^0$, it is congruent to $\angle DCB$, therefore $\triangle BDC$ is isosceles with $BD = CD = b$.

$AD < CD$ since AD lies against the lesser angle in $\triangle ADC$, i.e. $AD<b$, which means that $AB - BD < b$, or $l - b < b, \quad \Rightarrow \quad l < 2b$.

Thus, we have established that $b < l < 2b$ and the first step of the Euclidean algorithm can be written:

$$l = b + r; \quad r < b.$$

If b and l are commensurable, then, according to the algorithm (Th. 9.1.2), their greatest common measure will be the same as the GCM of the segments b and r.

Now the problem of finding the GCM of the base and a lateral side of an isosceles triangle has been reduced to finding the GCM of the segments b=AC and r=DA. AC is a lateral side, and DA is the base in an isosceles triangle ACD with a vertical angle of 36^0.

Thus the problem of finding the GCM of the base and a lateral side of an isosceles triangle with the vertical angle of 36^0 leads to the problem of finding the GCM of the base and a lateral side of another isosceles triangle with the vertical angle of 36^0. Similarly, this problem will lead to another problem of finding the GCM between the base and a lateral side for such a triangle, and thus the procedure of finding the required GCM is endless, i.e. the segments b and l are incommensurable, Q.E.D.

Remark. Let us notice that although we did use the angle measures in the formulation of this problem, we could formulate and solve it without resorting to them: instead of saying *an isosceles triangle with the vertical angle of* 36^0, we could say: *an isosceles triangle with the vertical angle equal to one half of a base angle.*

17. In an isosceles triangle with the measure of the vertical angle 108^0, each base angle (an angle whose side coincides with the base) will measure $\frac{1}{2}\left(180^0 - 108^0\right) = 36^0$.

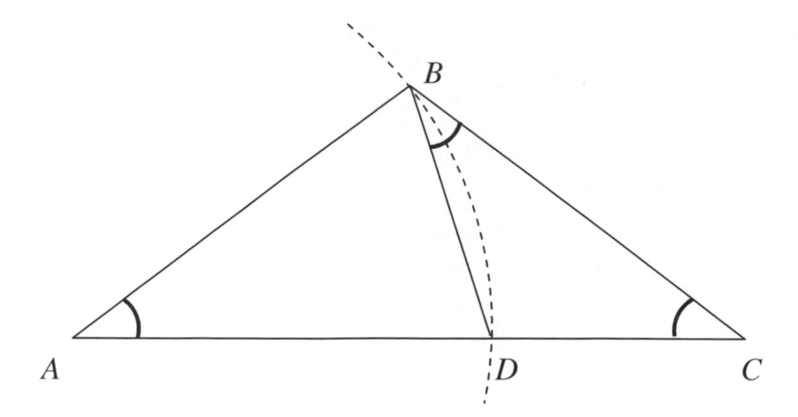

Let in ΔABC with $\angle ABC = 108^0$, $AB = BC = l; AC = b$. It should be noticed that AC is the greatest side of the triangle (why?), and according to the triangle inequality, $AC < AB + BC$; therefore, $l < b < 2l$.

Thus, we can write the first step of the Euclidean algorithm:

$$b = l + r; \quad r < l.$$

If b and l are commensurable, then, according to the algorithm (Th. 9.1.2), their greatest common measure will be the same as the GCM of the segments l and r.

In order to find the latter GCM, let us describe a circle of radius AB from A as a centre. Since $AC>AB$, the circle will cut AC at some point D. Then

$$DC = b - l = r.$$

Now the problem of finding the GCM of the base and a lateral side of an isosceles triangle has been reduced to finding the GCM of the segments $l=BC$ and $r=DC$.

It is easy to see that $\angle DBC = 36^0$, and therefore ΔCDB is an isosceles triangle with the vertical angle CDB measuring 108^0.

Thus the problem of finding the GCM of the base and a lateral side of an isosceles triangle with the vertical angle of 108^0 leads to the problem of finding the GCM of the base and a lateral side of another isosceles triangle with the vertical angle of 108^0. Similarly, this problem will lead to another problem of finding the GCM between the base and a lateral side for such a triangle (see the diagram below), and thus the procedure of finding the required GCM is endless, i.e. the segments b and l are incommensurable, Q.E.D.

Remark. Let us notice that although we did use the angle measures in the formulation of this problem, we could formulate and solve it without resorting to them: instead of saying *an isosceles triangle with the vertical angle of* 108^0, we could say: *an isosceles triangle with a base angle equal to one third of the vertical angle.*

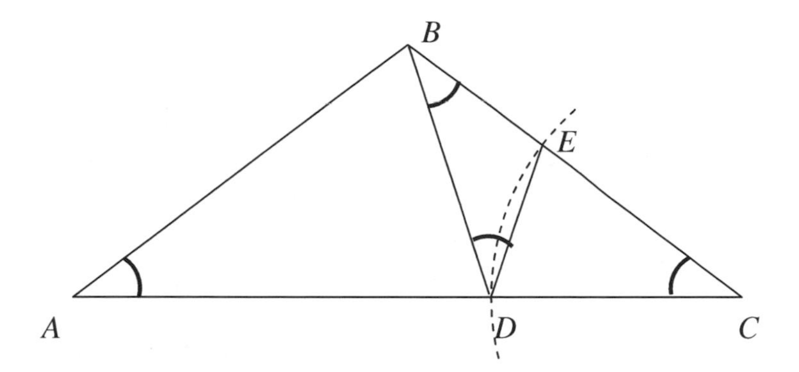

Sections 9.3-4

18. Hint: The triangles formed by the medians, the halves of the bases to which they are drawn and a pair of corresponding lateral sides, are similar by SAS; therefore the medians are in proportion with the lateral sides (or half-bases).

21. Let *AD* and *ME* be medians in triangles *ABC* and *MNP*, and $\triangle ADB \sim \triangle MEN$.

If we suggest in addition that the angles *B* and *N* (or the sides *AB* and *MN*) are corresponding, then it is easy to prove that $\triangle ABC \sim \triangle MNP$.

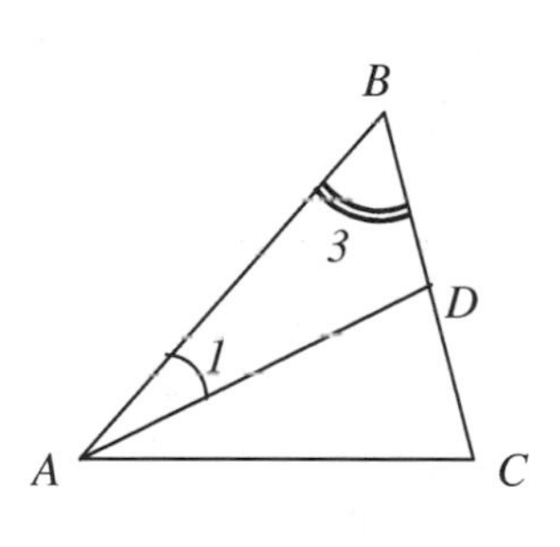

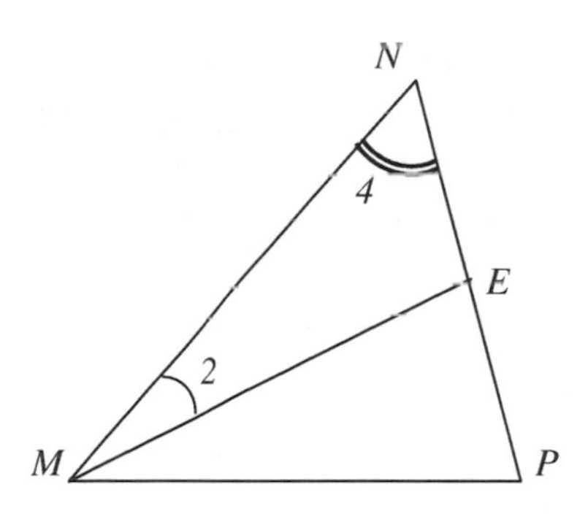

Since $\triangle ADB \sim \triangle MEN$, and the angles *B* and *N* are corresponding, $\angle B = \angle N$. Also, it follows from the similarity that

$$\frac{AB}{BD} = \frac{MN}{NE}; \quad \Rightarrow \quad \frac{AB}{\frac{1}{2}BC} = \frac{MN}{\frac{1}{2}NP}; \quad \Rightarrow \quad \frac{AB}{BC} = \frac{MN}{NP}.$$

Therefore $\triangle ABC \sim \triangle MNP$ by SAS.

In general case, however, if the angles *B* and *N* are not corresponding, the triangles *ABC* and *MNP* are not similar.

Suppose $\angle 1 \neq \angle 3$, and $\triangle ADB \sim \triangle MEN$ with $\angle B = \angle 2$; $\angle N = \angle 1$. We can show by construction that such a situation may take place, and the triangles *ABC* and *MNP* are not similar in this case.

Let ΔABC be given, and AD be a median in it.

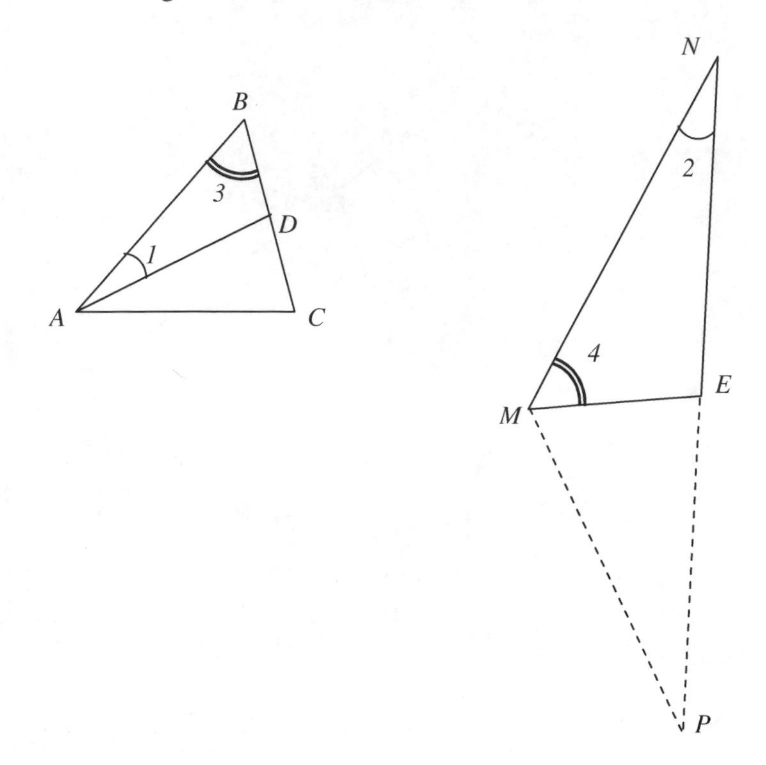

Construct $\angle 4 = \angle 3$ with its vertex at some point M. On one arm of this angle pick an arbitrary point N and construct $\angle 2 = \angle 1$ with its vertex at N and one arm lying on NM. Let the noncommon sides of the two constructed angles intersect at some point E (how do we know that they *will* intersect?). Thus we have constructed a triangle MNE that is similar to ΔABD by two angles (AA test).

Extend NE beyond E by a segment $EP=NE$. Draw PM. By construction, ME is a median in ΔMNP, and $\Delta ADB \sim \Delta MEN$.

At the same time ΔMNP is in general not similar to ΔABC: if, for instance, $\angle B$ is the greatest angle in ΔABC, then $\angle NMP > \angle 4 = \angle B$ is greater than any angle of ΔABC, thus the triangles ABC and MNP are not similar in such a case.

25.

CONSTRUCTION. Construct the given angle with its vertex at some point A. Lay off on its sides the segments $DB=2U$ and $DF=3U$, where U is an arbitrary segment. On BF from B, lay off a segment BC congruent to the given third side.

Through C draw a line parallel to DF; it will intersect (why?) the extension of BD beyond D at some point A. ΔABC is a sought for.

PROOF. ΔABC is similar to ΔDBF since their sides are parallel , and hence the corresponding angles are congruent. Therefore the ratio of the sides AB and AC is 2:3. Also, BC is congruent to the given third side by construction.

INVESTIGATION. The solution is unique since any other triangle satisfying these conditions will be congruent to ΔABC by ASA.

27. Let some segment a be a side of a sought for rhombus.

ANALYSIS. The diagonals in a rhombus are perpendicular (Th.7.2.2) and bisect each other; thus they divide the rhombus into 4 congruent right triangles. For the given rhombus, the ratio of the legs in such a triangle will be 2:3, the same as for the diagonals, since each leg is a half-diagonal.

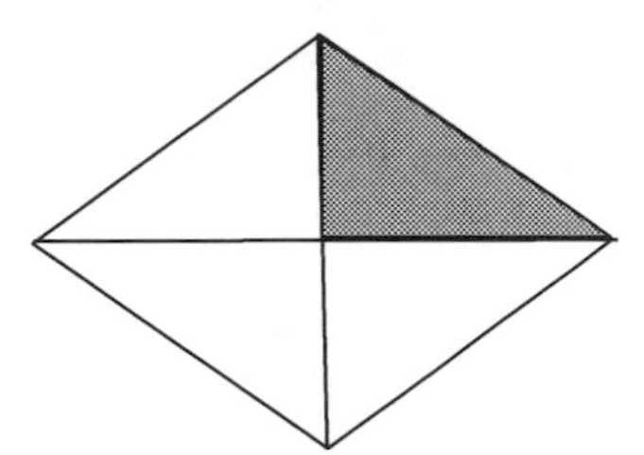

CONSTRUCTION. Let us construct a right triangle with the ratio of the legs 2:3 and the hypotenuse congruent to the given side.

In order to do this, we draw two mutually perpendicular lines that intersect at some point O and lay off from this point a segment OA congruent to $2u$ (where u is an arbitrary segment) on one line and a segment OB congruent to $3u$ on the other one. Join A and B. Thus we have constructed a triangle similar (by SAS) to the triangle that constitutes one-quarter of a sought for rhombus.

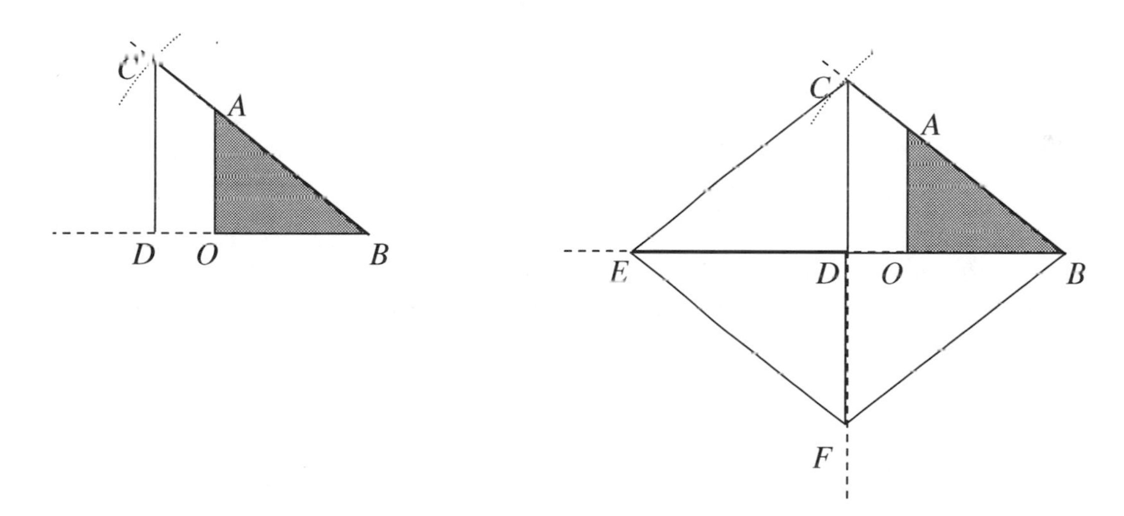

Extend BA beyond A and lay off on it $BC=a$, where a is the given side of a rhombus. From C draw a line parallel to OA till it intersects the extension of BO at some point D. Then, by construction, ΔBCD is similar to ΔBAO, hence $CD:DB$=2:3.

In order to complete ΔBCD into a rhombus with a side a, we should construct the triangles that are the symmetric images of BCD in the lines CD and BD. This can be done, for example, by extending BD beyond D by a segment $DE=DB$, by extending CD beyond D by $DF=CD$, and joining the points B with F, F with E, and E with C.

Let us prove that $BCEF$ is a sought for rhombus.

PROOF. The right triangles BDC, BDF, EDF, and EDC are congruent by two legs; thus $BF = FE = EC = CB = a$, and the figure is a rhombus with the given side. The ratio of the diagonals is $2CD$:$2DB$=2:3, which is the required ratio of the diagonals.

INVESTIGATION. Any right triangle with the ratio of the legs 3:4 will be similar to triangle AOB by two legs (or SAS), and therefore any such triangle will have an acute angle congruent to $\angle ABO$. Then any right triangle with the given ratio of its legs and the hypotenuse a will be congruent to ΔBCD by an acute angle and the hypotenuse. Therefore, any rhombus satisfying the given conditions will be congruent to *BCEF*, and the problem has a unique solution.

34. Hint. $\angle BCA = \angle ABP$ (why?), and therefore the triangles are similar by two angles.

43. Since a triangle has only three sides, and a square has exactly four vertices, it is inevitable that two vertices of an inscribed square will lie on one side of the triangle. Suppose ΔABC in the diagram below is the given triangle.
Construction. Let us inscribe a square with two vertices located on *AC* and the other two – on *AB* and *BC* respectively.

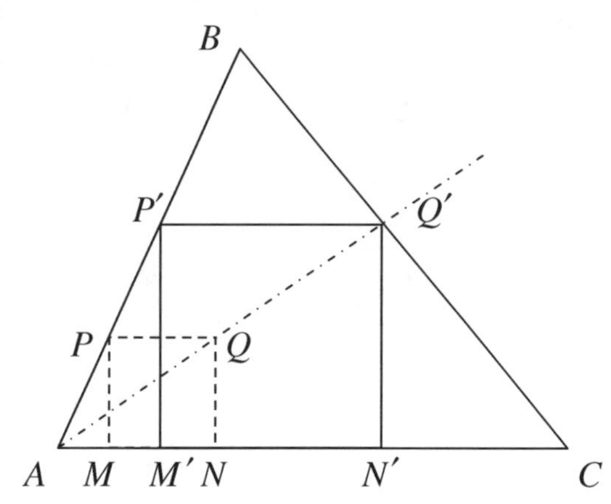

We shall start with drawing a square with one side lying on *AC* and one vertex located on *AB*:
from an arbitrary point *P* on *AB*, drop a perpendicular *PM* onto *AC* (*M* is the foot of the perpendicular);
from *M*, describe an arc of radius *MP* until it cuts *AC* at some point *N* between *M* and *C*;
through *P*, draw a line parallel to *AC*, and from *N* erect a perpendicular to *AC*; this perpendicular will intersect the above line (since the line is parallel to *AC*, it cannot be parallel to another line through *N*) at some point *Q*.

By construction, *PMNQ* is a rectangle; also *PM*=*MN*, hence it is a square.

From A, draw a ray through Q; it will intersect *BC* at some point Q'. Through Q', draw two lines: one of them // *AC* and the other one perpendicular to *AC*. They will cut *AB* and *AC* at some points that we shall name P' and N' respectively. From P', drop a perpendicular onto *AC*; let its foot be labeled M'. Let us show that $P'M'N'Q'$ is a square; then it will be a sought for square.
PROOF. By construction, $Q'N' \perp AC$ and $Q'P' // AC$; then by Corollary 1 of Th.6.2.2, $Q'N' \perp Q'P'$.

Also, by construction, $P'M' \perp AC$; $\Rightarrow P'M' \perp P'Q'$ (by the same corollary). Thus, $P'M'N'Q'$ is a rectangle. In order to show that it is a square it will be enough (why?) to prove that two of its neighbouring sides are congruent.

Let us show that $P'Q' = Q'N'$.

According to Lemma 9.2.1, $\Delta APQ \sim \Delta A'P'Q'$ and $\Delta ANQ \sim \Delta A'N'Q'$; therefore, $\dfrac{P'Q'}{PQ} = \dfrac{AQ'}{AQ}$ and $\dfrac{Q'N'}{QN} = \dfrac{AQ'}{AQ}$,

which entails $\dfrac{P'Q'}{PQ} = \dfrac{Q'N'}{QN}$; $\Rightarrow$ $\dfrac{P'Q'}{Q'N'} = \dfrac{PQ}{QN} = 1$; $\Rightarrow$ $P'Q' = Q'N'$.

Hence, $P'M'N'Q'$ is a square and it is a sought for figure, Q.E.D.

<u>INVESTIGATION.</u> Let us show that such a square is unique for each side of the triangle, i.e. there are exactly 3 distinct solutions, one for each side.

First of all let us show that the solution described above always exists.

There always exists a perpendicular from an exterior point (Q) onto a line (AC) (Th. 3.2.4), and it is unique (Corollary 2 from the exterior angle theorem). According to the compass postulate, one can lay off exactly one segment congruent to PM on ray MC. By Th. 3.2.3, exactly one perpendicular to AC can be erected from N, and by Th. 6.1.1, a line parallel to AC can be drawn through P. Such a line is unique according to the parallel postulate.

A line perpendicular to AC will intersect with a line parallel to AC (if we erect a perpendicular from some point on AC, it will be parallel to a perpendicular dropped onto AC; then if a perpendicular l drawn to AC is parallel to AC, it will contradict to the parallel postulate: there would be two lines: AC and a perpendicular erected from some point on AC, that are parallel to the same line l). A point of intersection (Q) is unique (AXIOM 1).

According to the same axiom, AQ can be drawn, and since it lies in the interior of $\angle BAC$, it will intersect BC at some point Q'. A perpendicular from Q' onto AC exists (Th.3.2.4) and it is unique (Corollary 2 to the exterior angle theorem).

It can be shown that since $\angle C$ is acute, this perpendicular will cut AC, not its extension: cutting the extension (at some point N' as in the diagram below) would contradict the exterior angle theorem for $\Delta Q'N'C$, since in this triangle its interior angle $\angle Q'N'C$ would be right whereas a non-adjacent exterior angle $\angle Q'CA$ would be acute.

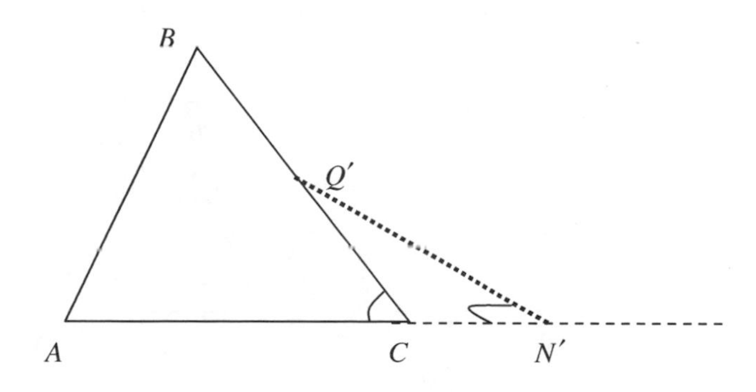

The rest of the construction is substantiated in the proof section of this solution. Thus the described above procedure results in the construction of exactly one sought for square for each side (we still have to prove that other squares cannot be obtained, maybe by a different constructing procedure).

It is interesting to notice that if one of the angle of the given triangle had been obtuse, the solution would not have existed for either side including that angle (why?).

Now let us show that there is no other square with two of its vertices on AC and the other two on AB and BC respectively, and therefore, the solution is unique for each side.

Suppose there is a square that is inscribed as required into the given triangle, with one of its sides lying on AC, and at least one of its vertices does not coincide with any of P', Q', N', M'.

All possible options are illustrated in the diagrams below, where the sides of the suggested "squares" that are different from $P'M'N'Q'$ are shown in dotted lines. As one can easily see, in cases when a side of a new square intersects a side (sides) of $P'M'N'Q'$, the suggested existence of a new square contradicts to the parallel postulate. In the only case when the sides of the new "square" are parallel to the respective sides of $P'M'N'Q'$ (lower right diagram), the sides of the new "square" cannot be congruent: the one that is perpendicular to AC is greater than $P'M'$ whereas the side lying on AC is less than $M'N'$.

Therefore, the constructed above square is the only one with two vertices on AC and the other two located on AB and BC respectively. Thus, the problem has exactly three solutions: one for each side.

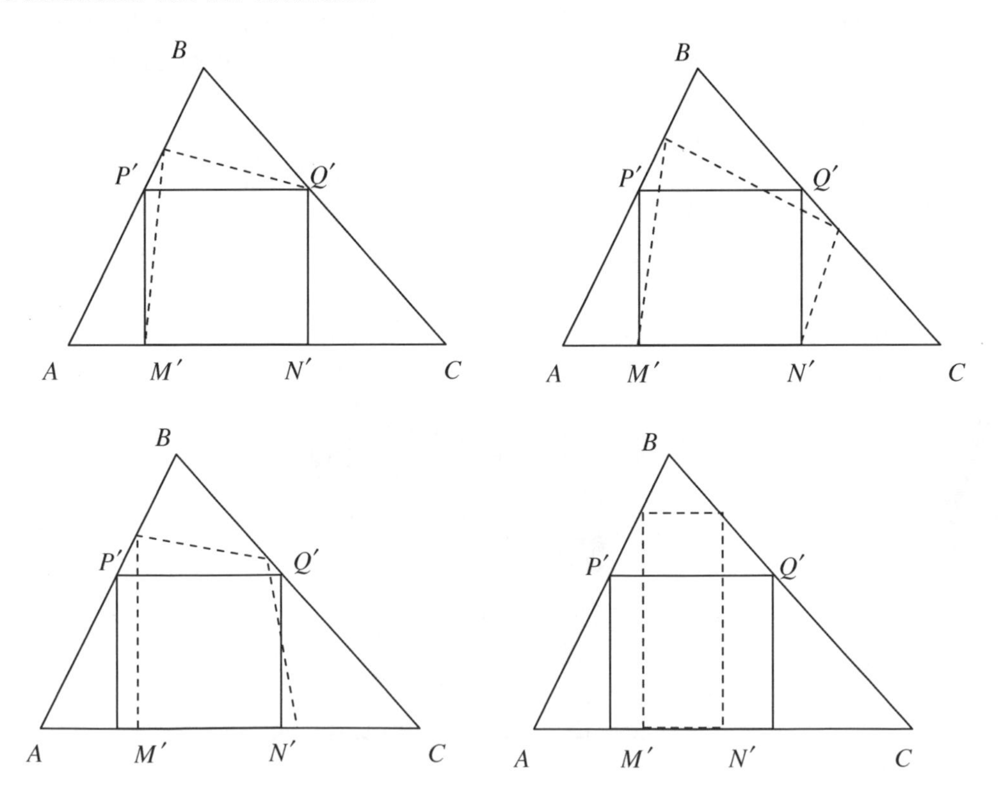

Section 9.5

16. There are two possible solutions since O may be an *internal* or *external* centre of similitude.

In the first case, when the centre is internal, the homothetic image of the polygon is presented by the polygon denoted **(i)** in the diagram. In this case, $OA' = AA' - OA = 3OA - OA = 2OA$, and the coefficient of similitude (magnification ratio) is equal to 2. Hence, $A'B' = 2AB = 12cm$.

In the second case, when O is an external centre of similitude, the homothetic image of the polygon is denoted **(e)** in the diagram. In this case $OA' = AA' + OA = 3OA + OA = 4OA$, and the magnification ratio is equal to 4, which implies $A'B' = 4AB = 24cm$.

No other possibilities exist: a homothetic image of A must lie on the line passing through O and A, and, according to Axiom 3, for each half-line emanating from A, there is exactly one endpoint A' of the segment AA' congruent to a given segment $3OA$. (There are two such half-lines: one of them passes through O, and another one is the extension of AO beyond A in the opposite from O direction).

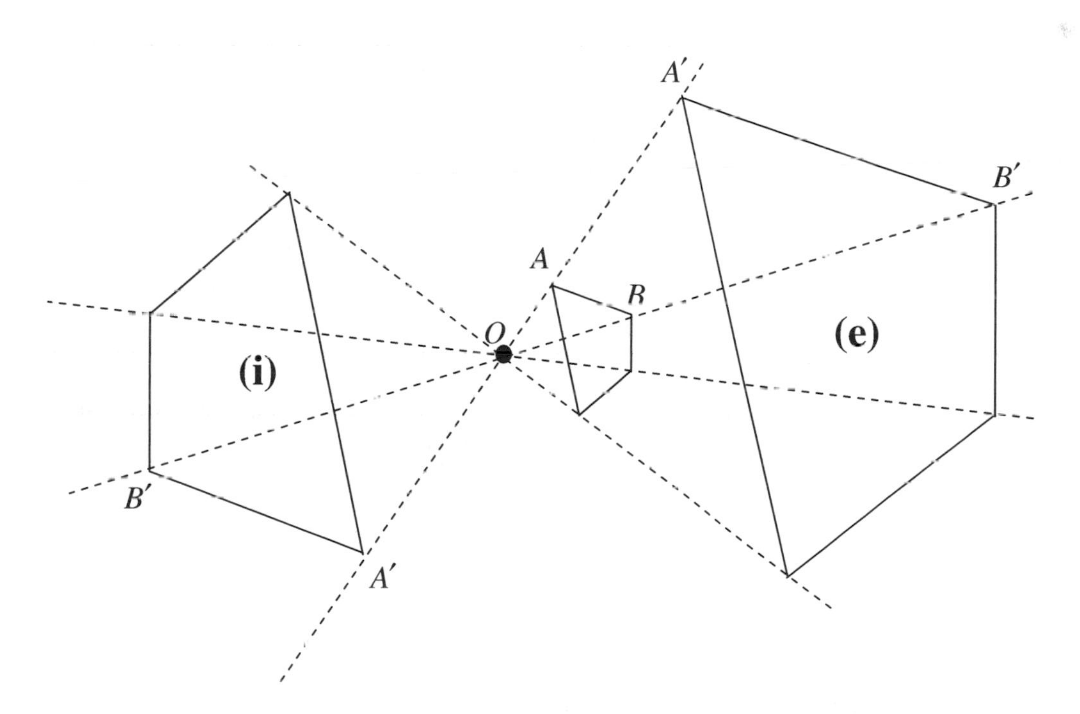

Section 9.7

4. SOLUTION (a brief description).

ANALYSIS. Any triangle with two angles congruent to the given two angles will be similar to a sought for triangle. Let us construct any such triangle and then apply a homothety that will rescale the triangle so as to make the inscribed circle of the given radius.

CONSTRUCTION. Construct a triangle ABC with the given two angles.

Inscribe a circle in that triangle; let its centre be at some point O, and its radius be r.

Apply to ΔABC a homothety with the centre of similitude at O and the magnification ratio equal to R/r, where R is the given radius of the inscribed circle. This can be done as follows: draw the three radii from O to the points of contact of the circle inscribed in ΔABC with its sides and extend them into segments OD, OE, and OF, each of them congruent to R. Through D, E, and F, draw the lines tangent to the circle. Each of these lines will be parallel to a respective side of ΔABC (why?), hence they will intersect (why?) and thus form a triangle, which is similar to ΔABC (why?).

<u>PROOF</u> This is a sought for triangle, since its angles are congruent to the corresponding angles of ΔABC, and the radius of the inscribed angle is R.

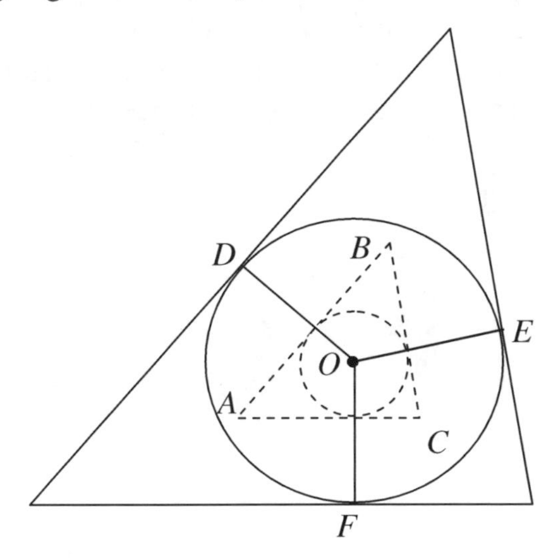

<u>INVESTIGATION.</u> It is easy to show that all triangles that satisfy the given conditions are congruent, and therefore, the solution is unique.

Section 9.9

16. The segments are proportional to the respective legs (Th. 9.8.2), and their sum is congruent to the hypotenuse $c = \sqrt{a^2 + b^2}$.

Hence the segments will be $\dfrac{a}{a+b}\sqrt{a^2+b^2}$; $\dfrac{b}{a+b}\sqrt{a^2+b^2}$.

22. The possible cases are shown in the diagrams below. In each case the lateral side is a mean proportional between the hypotenuse and its projection onto the hypotenuse in a right triangle formed by the lateral side as one leg, diameter BE, and the chord joining the respective base vertex of the triangle with an endpoint of the diameter.

Thus in the first case (the vertical angle is acute),
$BC = \sqrt{BD \cdot BE} = \sqrt{(R+OD)(R+R)} = \sqrt{(5+4)(5+5)} = 3\sqrt{10}$.

In the second case (the vertical angle is obtuse),
$BC = \sqrt{BD \cdot BE} = \sqrt{(R-OD)(R+R)} = \sqrt{(5-4)(5+5)} = \sqrt{10}$.

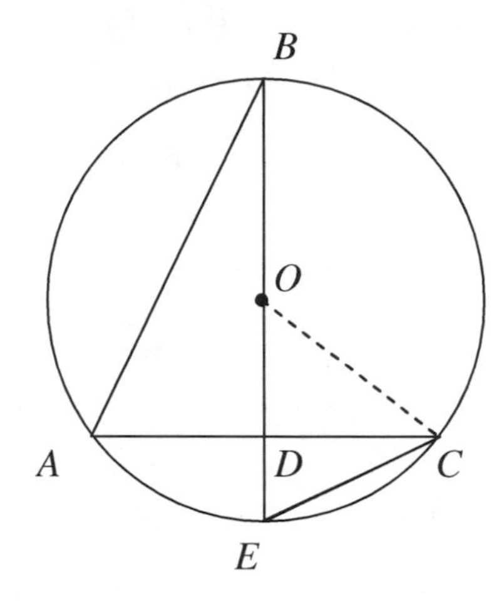

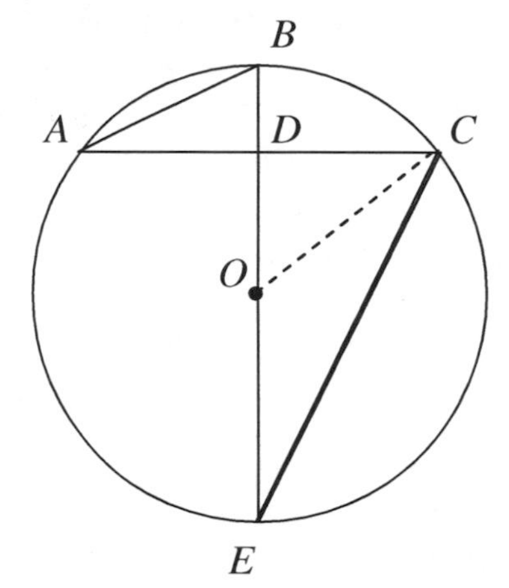

Let us notice that the same results can be obtained by applying the Pythagorean theorem twice: to the triangles *ODC* and *BDC*.

Section 9.12

6. A possible solution follows. $a\sqrt{5} = a\sqrt{1+4} = a\sqrt{1^2+2^2} = \sqrt{a^2+(2a)^2}$.

Thus the segment can be constructed as the hypotenuse in a right triangle with the legs a and $2a$.

9. Review

16. Let A be the given point on a circle (O, r) – a circle of radius r with its centre at O. If M is the midpoint of some chord AB passing through A, then M is the foot of the perpendicular dropped from O onto AB (Th. 8.1.3). In a circle for which OA is a diameter, angle OMA is an inscribed angle standing on that diameter; hence our hypothesis: The sought for locus is the circle for which OA is a diameter. This locus is shown in a dotted line in the diagram below.

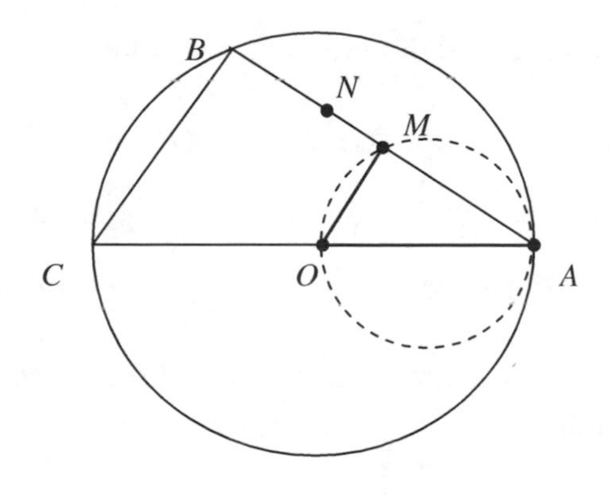

Really, let M be a point on this (dashed line) circle. Draw a chord (AB) through A and M. Angle OMA is right as an inscribed angle standing on a diameter, hence by Theorem 8.1.2, M is the midpoint of AB. (An alternate proof can be based on the similarity of the triangles CBA and OMA). Thus every point M lying on the circle for which OA is a diameter does bisect the corresponding chord.

Now we have to show that any point N that is not located on this circle is not the midpoint of the chord passing through it and point A. This follows immediately form Axiom 3.

Therefore the circle for which OA is a diameter is a sought for locus.

32. Let the ratio $m:n$, the base b, and a circle be given.

ANALYSIS. Let $\triangle ABC$ in the diagram below be a sought for triangle, with $\frac{AB}{BC}=\frac{m}{n}$, and $AC=b$. The bisector BD of $\angle ABC$ will divide AC in the ratio $m:n$ (Th.9.8.3). Also, the extension of this bisector will divide the arc "beneath" AC (i.e. lying on the opposite from B side of AC) into congruent parts, since the respective inscribed angles are congruent.

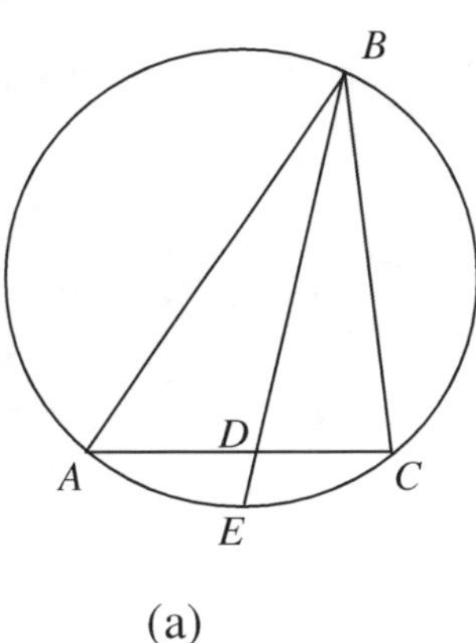

(a)

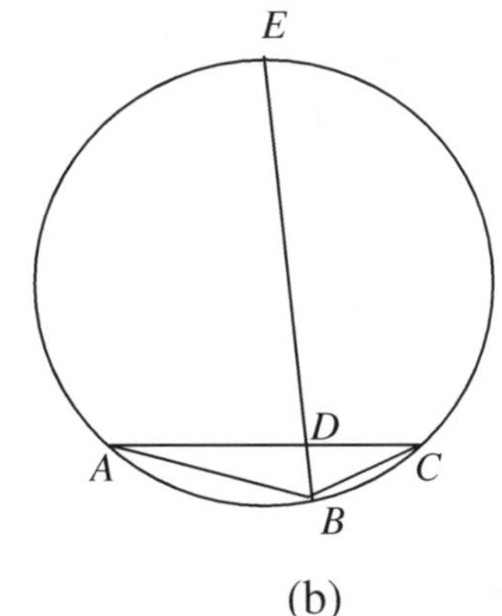

(b)

CONSTRUCTION. From an arbitrary point A on the circle, describe an arc of radius b; it will cut the circle at two points. Let C be one of these points. Draw AC and divide it in the ratio $m:n$ (a standard problem): $\frac{AD}{DC}=\frac{m}{n}$. Find the midpoint E of one of the arcs AC (we have chosen the smaller arc on diagram (a) and the greater one on diagram (b)).

Draw ED and extend it till it intersects the circle at some point B.

$\triangle ABC$ is a sought for triangle.

SYNTHESIS. BD is a bisector by construction: angles ABE and CBE are inscribed in congruent arcs. It follows from Theorem 9.8.3. the ratio $\frac{AB}{BC}$ is equal to the ratio $\frac{AD}{DC}$

Thus $\triangle ABC$ is inscribed in the given circle and has the given base and ratio of the lateral sides, Q.E.D.

INVESTIGATION. In general, there are exactly two solutions. Prove that any other triangle satisfying the given conditions will be congruent to one of the triangles constructed: triangle (a) with $\angle ABC$ acute or (b) with $\angle ABC$ obtuse, as shown in the above diagram.

In the particular case when the base is congruent to the diameter of the circle, the two triangles represented by (a) and (b) will be congruent, hence the solution will be unique (It will be a right triangle in this case). In any other case the problem has exactly two solutions.

33. Let A be the given point on a circle (O, r) – a circle of radius r with its centre at O, segment b the given base, and m the given median.

ANALYSIS. Let $\triangle ABC$ be a sought for triangle, with the base $AC = b$ and median $AM = m$.

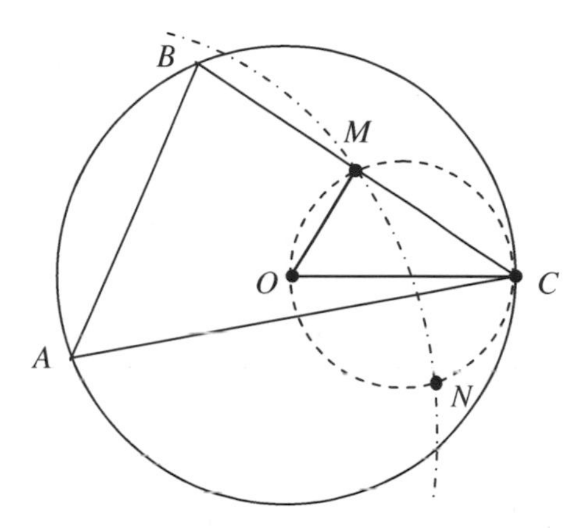

According to the result of Problem 16 of this section, the midpoint M of the side BC of the sought for triangle must be located on the circle that has OC as a diameter. Also, M must lie at the distance m from point A, hence it must lie on the circle of radius m described from A as a centre. Thus point M can be found as a point of intersection of the two loci: a circle with diameter OC, and a circle of radius m with the centre at A.

CONSTRUCTION. From an arbitrary point A on the given circle, describe a circle of radius b. If $b > 2r$, the circle will not touch or intersect the given circle, and the problem will have no solutions; otherwise the two circles will have at least one point C in common, and $AC=b$ by construction.

Draw a circle Γ with diameter OC (bisect OC and describe a circle of radius $\frac{1}{2}r$ from its midpoint). From A, describe a circle Λ of radius m. If the circles Γ and Λ have a common point M, draw a chord from C through M; it will cut the circle at some point B.

$\triangle ABC$ is inscribed into the given circle, it has a base $AC=b$ and median $AM=m$ by construction, hence it is a sought for circle.

PROOF. It is included in the CONSTRUCTION (see the latter paragraph).

INVESTIGATION. The problem may have up to two solutions, since the loci (two circles may have up to two points of intersection (M and N on the diagram) if $b < 2r$. If $b= 2r$, AC will be a diameter, and the two solutions will be congruent triangles.

39. Construct a triangle having given two sides and the bisector of the angle included between the given sides.

Let b, c, and l be the given sides and the given bisector respectively.

<u>ANALYSIS.</u> Let ΔABC with the sides $AC = b$, $AB = c$, and the bisector $AO = l$ from the vertex A be a sought for triangle ($\angle 1 = \angle 2$ since AO is a bisector).

Through C, draw a line parallel to AB. This line cannot be parallel to AO (since AO and AB intersect – they are not parallel), hence it will intersect the extension of AO (why the extension, and not AO itself?) at some point D.

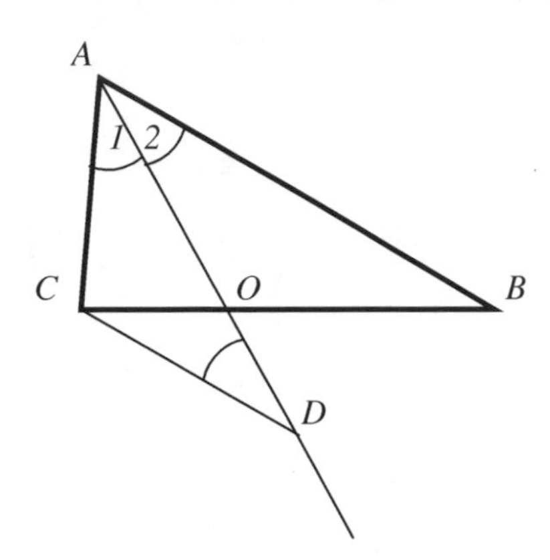

$\angle COD = \angle 2$ as alternate angles formed by parallel lines CD and AB and their transversal AD. Hence $\angle COD = \angle 1$, and ΔDCA is isosceles with $CD=CA=b$.

$\Delta AOB \sim \Delta COD$ by two angles (AOB and COD are verticals), therefore

$\dfrac{OD}{AO} = \dfrac{CD}{AB} \Rightarrow \dfrac{OD}{l} = \dfrac{b}{c} \Rightarrow OD = \dfrac{b}{c}l$; hence $AD = l + OD = l + \dfrac{b}{c}l$.

Thus we can determine and construct AD, which means we know all the sides in triangle ACD, and the latter triangle can be constructed.

<u>CONSTRUCTION.</u> Construct $OD = \dfrac{b}{c}l$ as the fourth proportional (a standard construction) and extend it by l to obtain $AD = l + OD = l + \dfrac{b}{c}l$.

Construct ΔACD by SSS.

Construct $\angle 2$ congruent to $\angle 1$, with one side coinciding with AO and the other side m lying on the opposite from AC side of AO.

Extend AO beyond O until it intersects m at some point B (explain why it will intersect m).

ΔABC is a sought for triangle.

<u>PROOF.</u> It follows from the construction.

<u>INVESTIGATION.</u> The solution is unique. Prove that any other triangle with the sides congruent b and c and the included between them bisector congruent to l will be congruent to the triangle constructed. (Hint: Use SSS and SAS tests or superposition).

42. <u>Hints.</u> Let a be the segment of the tangent from the given point to the circle; it can be constructed by drawing the tangent line. Let b and c be respectively the exterior and interior segments of the secant from the given point, and let their ratio be $c:b=m:n$, where m and n are known numbers (they also may represents the lengths of given segments).

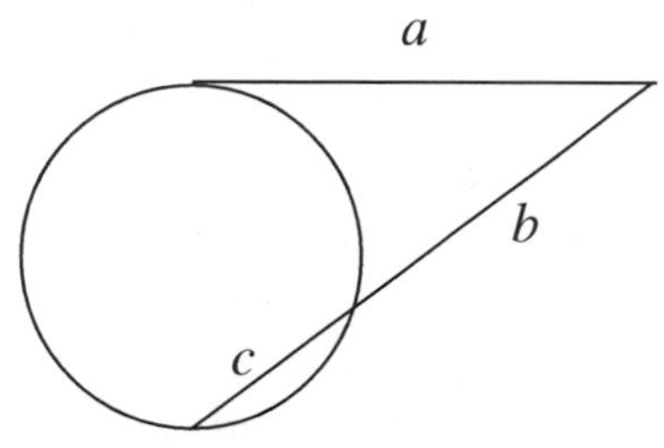

Then, it follows from Th.9.10.2 that $b(b+c)=a^2;\quad \Rightarrow \quad b(b+\frac{n}{m}b)=a^2$. Then we can express b in terms of a: $b^2=\frac{m}{m+n}a^2;\quad \Rightarrow \quad b=\sqrt{\left(\frac{m}{m+n}a\right)a}$.

Thus segment b can be constructed as the mean proportional between the segments a and $x=\frac{m}{m+n}a$. The first of these segments is known, and the second can be constructed in a standard manner by cutting the sides of an angle with parallel lines (finding the fourth proportional).

Then b can be constructed, for instance, as the altitude onto the hypotenuse in a right triangle with the projections of the legs onto the hypotenuse equal a and x.

Summary and Comments to Chapter 1.

As soon as we say that we are going to study Euclidean geometry, the following two questions emerge:

(1) What is this geometry about?

(2) How are we going to study it?

Section 1.1 prepares students to the discussion concerning the first question. They are asked to propose answers to that question, based on their intuitive ideas of what is geometry. Their proposals should be discussed (An example of such a discussion is provided in the **Supplement to Section 1.1**, following below these comments).

As a result of the discussion, students should

(i) agree upon which properties are not geometrical (e.g. physical, chemical, etc.);

(ii) be aware that the essence of a geometry lies in the definition of what are the geometrical properties in the geometry;

(iii) understand, at intuitive level, that the essence of Euclidean geometry (the one we are going to study) is expressed in the axiom (common notion) proposed by Euclid for comparing geometric figures: *The things are equal if they coincide.* (See Problem 3 to section 1.1).

The rest of the discussion of what are the geometric properties of Euclidean geometry is postponed till section 1.3, where these will be defined.

Also, in **section 1.1** we mention that although geometry emerged as a practical art based on measurements, we are not going to study it *empirically*: our results and conclusions will not be based directly on observations and measurements. Instead, our knowledge will be derived *theoretically*. The latter term should be explained intuitively and supported by examples, e.g. from Physics.

The exact meaning of the term *theoretically* is clarified in **section 1.2**: we shall derive our results by means of logic from a few basic statements (axioms or postulates) assumed to be true. The ideas of theoretical studies based on axioms and logic are new for students, that is why it is important to discuss in detail all the problems of the section.

All the problems of the section are provided with solutions (at the end of the book or in the additional solutions for the instructor's manual), and some of these constitute part of the discussion. For example, students should read the solution to problem 14, discussing the definitions (they should do this, of course, after a guided discussion of their own). There will be more problems and materials discussing the notions of axioms, reasoning, logical substantiation and deduction in the sections 1.3 and 2.1.

In **section 1.3** the undefined notions (primitive terms) and the first axioms are introduced. These axioms define the subject of Euclidean geometry: a geometry that studies properties invariant under rigid motions (isometries).

The set of axioms is approximately equivalent to the first four Euclidean postulates, with the difference in having axiomatic description of rigid motions versus no description at all in *The Elements*. It is recommended that the instructor read "Note on Common Notion 4" in the English translation of "Euclid's Elements" by Sir Thomas Heath (pages 224-228, vol. 1, in the second edition by Dover Publications). The note discusses the role of the notion of congruence, which has not been either discussed or postulated in "the Elements". As follows from that discussion, supplied with historical remarks, ancient scholars did have a feeling that something else should be done and therefore were "shy" of using the rigid motions for proofs in many instances. (See also the **Supplement to Section 1.1** below these comments).

After the existence of rigid motions (isometries) has been postulated, it is important that students learn how to use them. That is why problems #34-43 are the essential problems of the section (all of them are supplied with solutions, but students should try their best first).

Problems 14-15 and 23-25 in section 1.3 are important as well, since they give students an opportunity to apply deductive thinking in simple situations and to develop the right language for explaining and substantiating their solutions.

The additional material on transformations and equivalence relations (and problems 44-56) is beneficial, especially for advanced students, but not essential for the course.

<u>Comments to section 1.1.</u>

In this section we make our first attempt to determine the subject of our studies.

It is a good idea to discuss with the students, what is geometry about from their point of view. What kind of properties of objects do we study? Are physical, chemical properties important? – Probably not. As soon as we agree on that, we understand that in geometry we are not going to deal with real, or physical, objects. We shall consider their idealizations, which we shall call "geometric figures".

Then, which properties of these are important? – This is another good topic for a discussion with students. Propose them to draw a few figures that they assume to have the same and different geometric properties. For example, consider the objects shown in Figure C1-1.1 below: three circles, three quadrilaterals, and five triangles. Which of these have the same geometric properties from your point of view? Which of these would you identify as identical (i.e. not different) from the viewpoint of our geometry (whatever we intuitively perceive as "our geometry")?

In order to direct the discussion and give students a hint, the instructor can ask the following question: What kind of experiments would you propose in order to answer the latter question? (This is really crucial!)

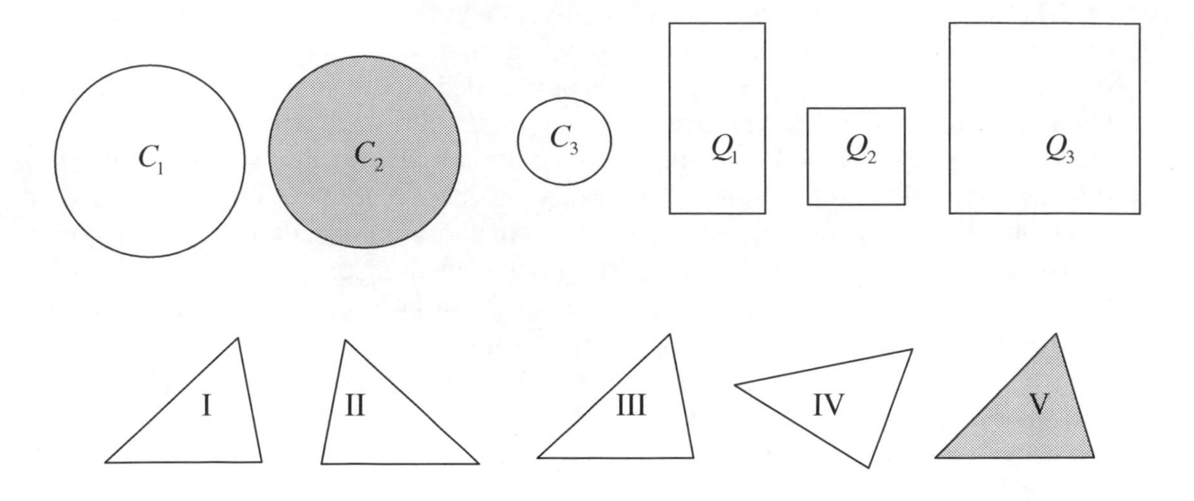

Figure C1-1.1

Usually students provide the following answers:
"Two figures are identical if they have the same sizes and shapes."
Then the following questions emerge:

(1) How are we going to measure the figures?
(2) What do we mean by *shapes*?

The second question is especially difficult to answer unless we assume that *shapes* are something we perceive intuitively and this notion cannot be explained in formal terms.

However, even the first question is not easy to answer. How are we going to measure figures? Suppose we have measuring tools such as a compass (for straight segments) and protractor (for angles); still it is not obvious which segments and angles are to be measured.

In the case of triangles, the simplest rectilinear figures (i.e. figures formed by straight segments), we can propose to measure all the sides and angles and then assume the rule: if each angle of one triangle has the same measure as an angle of another triangle, and the sides opposed these angles correspond to the same stretch between the legs of a compass, then the triangles are identical, from the viewpoint of Euclidean geometry.

Such a definition seems to be acceptable, however even in such a simple situation, we understand that we can measure sides and angles only within a certain accuracy. Thus the proposed criterion is not absolutely precise. For example, if we assume the accuracy of 0.05mm (or 0.02 inch), all the triangles in the above figure are identical, whereas if the required accuracy is 0.025mm (or 0.01 inch), the coloured triangle (triangle V) is not identical to the others. As for the triangles I-IV, they have been drawn by means of the drawing tools available at Microsoft Word software on the grid with the resolution 0.01 inch or 0.025 mm, so we cannot know whether they are identical within a finer accuracy.

As for non-rectilinear figures, even as simple as circles, it is extremely difficult to "measure" them (whatever this word may mean), not mentioning the accuracy problem. Some participants of the discussion would propose: measure the diameters! – it turns out we can only guess what would be diametrically opposed points (unless we use some results of Euclidean geometry), which makes our accuracy inherently flawed.

As for more sophisticated figures, for instance the ones shown below, what measurements should we do in such a case? Should we measure the distances between each pair of points of the figure? – This method is not viable, since it would require infinitely many measurements!

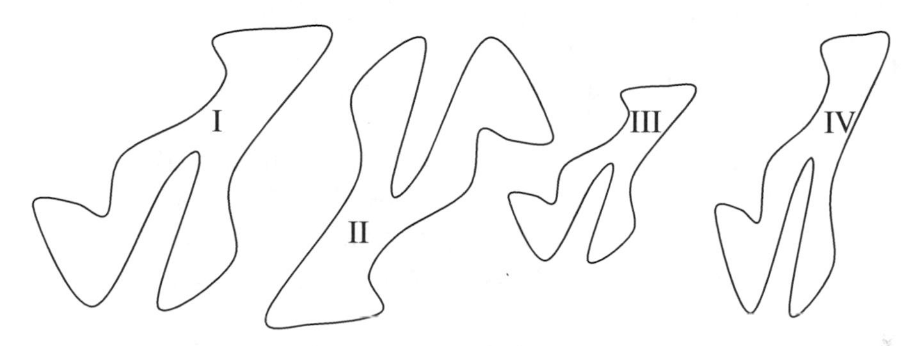

Figure C1-1.2

Thus it is not very easy to propose a good criterion to compare geometrical figures. Still, from a practical point of view we seem to have an idea how to compare figures, even sophisticated ones like in Figure C1-1.2: If we cut off from the sheet of paper, on which the figures are drawn, figure II, and by means of moving and rotating it we *impose* it exactly on figure I, and it will take exactly the same position on the sheet of paper, we can say that these two figures are identical. Such a method allows us to compare figures of any shapes without doing any measurements (whatever this word means)!

This method of comparing figures and identifying them was proposed by Euclid in *The Elements*. He formulated it Common Notion 4:

Things which coincide with one another are equal to one another.

It is clear from the proofs of the first results in *The Elements* that Euclid actually meant that one can move one figure and superimpose it onto the other in order to make the figures coincide. That is why in more modern formulations it is often stated as the following principle, usually called the *congruence (equality) by superposition*:

Two figures are equal (congruent) if one of them can be (super) imposed onto another so that they completely coincide in all their parts.

The above *superposition principle* states that two figures are considered to be identical if one of them can be moved so as to occupy exactly the same position in space (or in a plane) as the other. Thus, if we follow that principle, we assume that a motion without deformations, a so-called *rigid motion*, which is applied to a figure in order to compare it with another figure, *does not change the geometrical properties* of figures.

Now we can tell what the subject of Euclidean Geometry is:

Euclidean Geometry is the field of knowledge that studies those properties of geometrical figures that do not change if the figures are subjected to rigid motions.

One can be almost happy with this definition, and still it has a flaw: we define the geometrical properties through the notion of a rigid motion, or a motion without deformations, which is a physical notion.

This contradicts to our original idea of abstraction from the physical nature of objects in order to study the so-called idealizations of physical, or material, objects. Euclid and his followers seemed to be aware of this flaw; otherwise it is hard to explain why the superposition principle was used by Euclid only in the proofs of the first two criteria of equality of triangles and never again (even though it could be successfully employed in many other proofs).

Later in the course (in section 1.3) we shall define the notion of a *rigid motion* (*isometry*) rigorously, that is without referring to the physical properties of rigid bodies. Also, the existence of essential rigid motions (as mathematical objects) will be postulated and used for the proofs of the basic results.

Test to Chapter 1.

1. Quadrilateral *ABCD*, shown in the figure below, is congruent to each of the quadrilaterals shown in figures (a), (b), and (c). For each of these three quadrilaterals, propose a sequence of isometries that will transform *ABCD* into it. (You can only use isometries from Axiom 4 since the existence of other isometries or isometries with some additional properties has not been postulated). In each case try to find the solution that includes the use of as few steps as possible. (For convenience, label the vertices of each of the three figures).

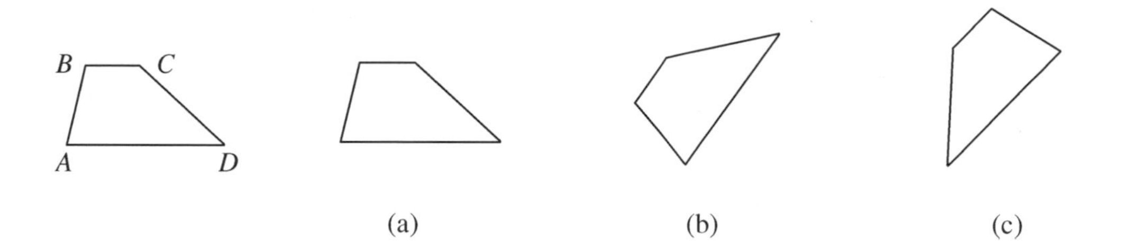

(a) (b) (c)

2. It is given that quadrilateral *ABCD*, shown in the figure below, is *symmetric in the line* passing through the midpoints *E* and *F* of its sides, which means the quadrilateral does not change under type (iii) isometry that leaves *EF* unchanged.
   Also, *ABCD* is congruent to each of the quadrilaterals shown in figures (a), (b), and (c). For each of these three quadrilaterals, propose a sequence of isometries that will transform *ABCD* into it. Propose a few solutions. In each case try to find the solution that includes the use of as few steps as possible. Can you avoid using certain types of isometries? Which one(s)?

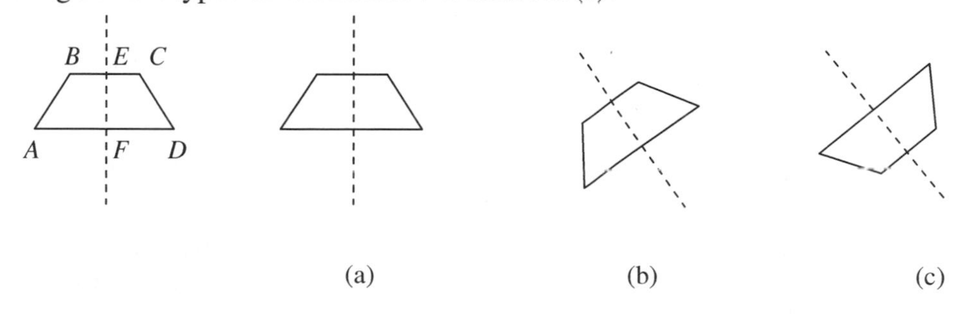

(a) (b) (c)

3. Prove that the radii of congruent circles are congruent.

Solutions to Chapter 1 Test.

1. Quadrilateral *ABCD*, shown in the figure below, is congruent to each of the quadrilaterals shown in figures (a), (b), and (c). For each of these three quadrilaterals, propose a sequence of isometries that will transform *ABCD* into it. (You can only use isometries from Axiom 4 since the existence of other isometries or isometries with some additional properties has not been postulated). In each case try to find the solution that includes the use of as few steps as possible.

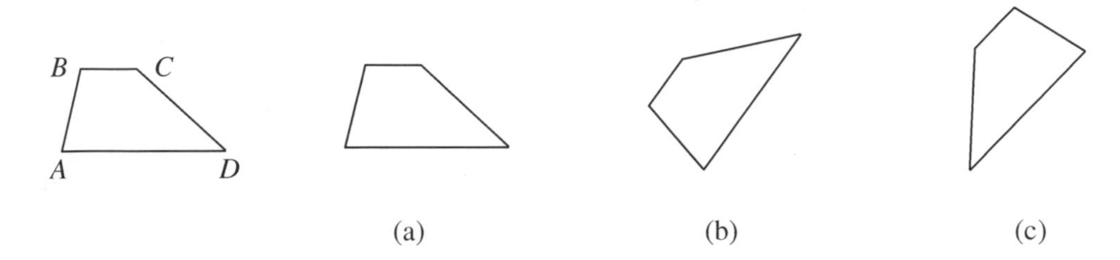

Solutions to Problem 1.

It is given that *ABCD* is congruent to each of the three figures; therefore for each of them there exists an isometry that can *superimpose ABCD* on the corresponding figure so that they will coincide *in all their parts.*

The latter means in particular that each side of *ABCD* will fall on its congruent, and each angle of *ABCD* will fall on a congruent angle. Such angles and sides that coincide when two congruent figures are made to coincide by an isometry, are called *corresponding.*

(a) Let us denote the quadrilateral in position (a) *MNPQ*. Since it is given that *ABCD* is congruent to *MNPQ*, which means they can be made to coincide as a result of a isometries, we can suggest that *MQ* is the side that is congruent to *AD* (i.e. they will coincide with one another when the first quadrilateral is superimposed onto the second one), *PQ* is congruent to *CD*, etc. Such pairs of sides or other elements (for instance, angles) of congruent figures that are respectively congruent, are called *corresponding.*

First step: Apply an isometry that carries point *A* into *M* (Type (i) – a translation of a point into a point).

As a result of this isometry, *ABCD* will take some position similar to the one shown in the figure below in dotted lines. (Notice that we cannot be sure that as a result of the first isometry the segment MD' (the *image* of *AD*) will fall exactly onto *MQ*: it is not guaranteed by a type (i) isometry.)

Second step: Apply an isometry that will make the ray emanating from *M* and passing through D' fall onto the ray emanating from *M* and passing through *Q* (Type (ii) – a rotation of a ray about a point).

Then MD', which is congruent to *AD*, and hence to *MQ*, will fall exactly onto *MQ*, and the quadrilaterals will coincide.

Remark. A similar solution could be carried out in many different ways, for example, by moving B onto N and then rotating NC' (the image of BC) so as to make it coincide with NP, as shown in the rightmost figure below.

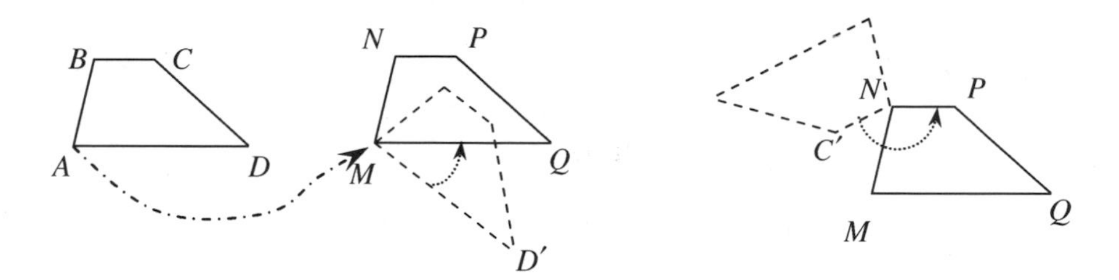

(b) We can solve this problem in a similar way (see the diagram below).

First step. We shall use an isometry of type (i) to move one of the vertices, for instance vertex D, so that it will coincide with the *corresponding* vertex Q. (We shall call two points in congruent figures *corresponding* if we expect them to coincide as a result of an isometry that makes the figures to coincide. If such points are vertices, then the pairs of *corresponding sides* meet in them. In our case, D and Q are corresponding vertices: AD and CD meet in D, and $A'Q$ and $C'Q$ meet in Q, and these are pairs of corresponding sides: $AD = A'Q$, and $CD = C'Q$)

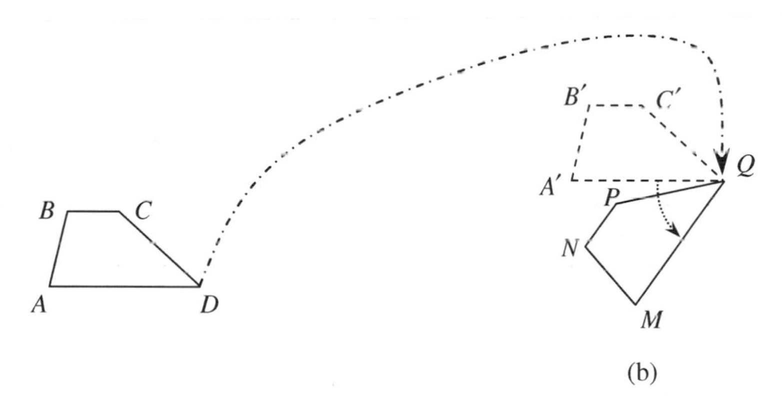

(b)

Second step. We shall apply an isometry of the second type that will rotate the ray emanating from Q and passing through A' until it coincides with the ray emanating from Q and passing through M. Then, since it is assumed that $DA=QM$ as corresponding sides, QA', which is congruent to DA (why?), will fall exactly onto QM (by Axiom 3). Then the two figures will coincide.

(c) In this problem we shall have to use a type (iii) isometry, since the $ABCD$ and the figure ($EFGH$) shown in diagram (c) have *different orientations*. Let us explain what it means.

As we can observe in the diagram below, the shortest rotation that will carry DC onto DA is to be directed counterclockwise, whereas the shortest rotation imposing EF (the side corresponding to DC) onto EH (corresponding to DA) is to be directed clockwise.

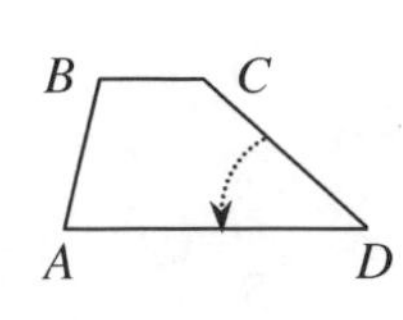

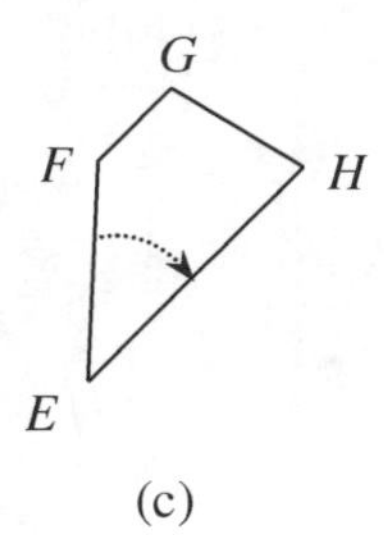

(c)

Because of the different orientations of the figures we shall have to make at least three steps in order to superimpose one onto the other.

First step. Translation: $D \rightarrow E$ (Type (i)).

Second step. Rotation: $EC' \rightarrow EF$ (Type (ii)).

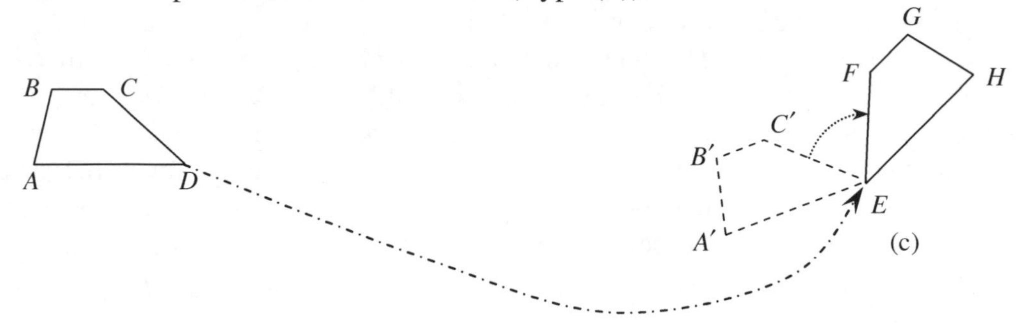

(c)

As a result of the first two steps, we shall transform *ABCD* into $A''B''FE$ shown in the figure below. Then we shall have to perform a type (iii) isometry, which will change the orientation of the figure.

Third step. Reflection in the line containing segment *EF* (Type (iii)).

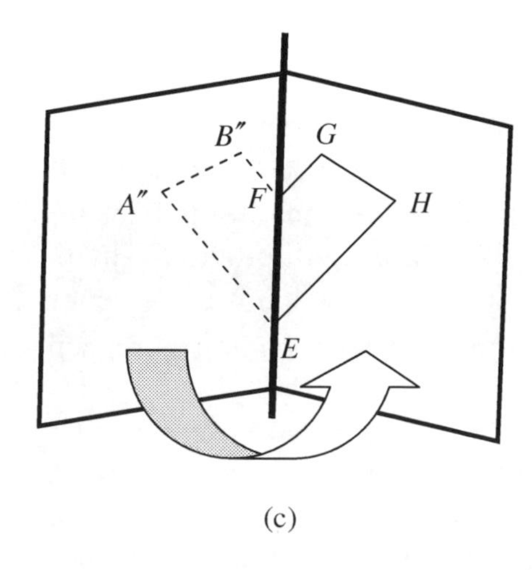

(c)

The latter isometry maps the half-plane that contains $A''B''FE$ onto the half-plane lying on the other side of *EF*, as shown in the figure above.

Then due to the congruence of the corresponding angles EFB'' and *EFG*, FB'' will fall onto *FG*, B'' will fall exactly onto *G* since the $FB'' = FG$, $B''A''$ will fall exactly onto *GH*, etc.

This step completes the transformation of *ABCD* into *HGFE*.

Remark. Had we known how to *bisect* (divide into two congruent parts) straight segments and angles, we could have used reflections (type (iii) isometries) more efficiently for solving the problems (a), (b), and (c).

For instance, we could have replaced the second and third steps of the solution of problem (c) by a single step: a reflection of the half-plane containing $A'B'C'E$ in the line that passes through *E* and bisects the angle between EC' and *EF* (the line is shown in bold in the figure below).

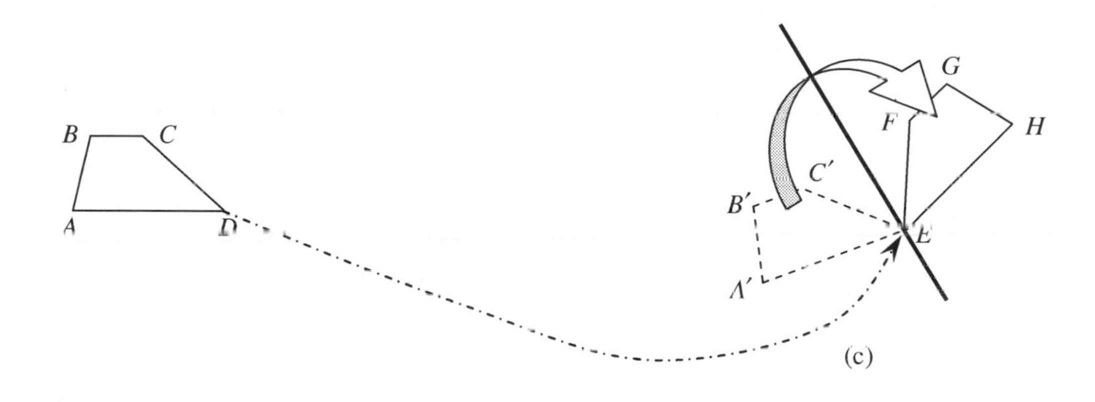

Reflections could be used for solving (a) and (b) as well.
For example, the diagram below illustrates a possible solution of (b).

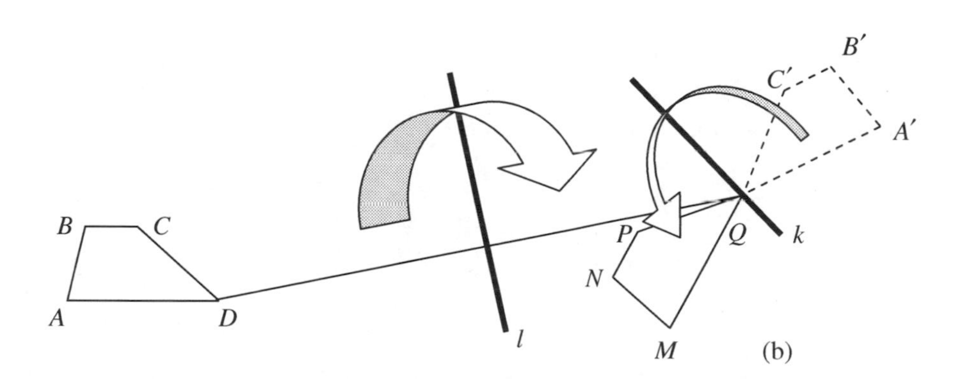

The transformation consists of two reflections. The first one reflects in line *l*, which bisects the segment *DQ* and makes congruent angles with the parts of this segment that lie in the two half-planes on the opposite sides of *l*. As a result of this reflection, $ABCD$ is transformed in $A'B'C'Q$. The second transformation is a reflection in line k that makes congruent angles with the segments QM and QA'.

Let us emphasize again that **such solutions cannot be justified** until we have shown how to bisect segments and angles. We shall consider such constructions in chapters 4 and 5.

2. It is given that quadrilateral $ABCD$, shown in the figure below, is *symmetric in the line* passing through the midpoints E and F of its sides, which means the quadrilateral does not change under type (iii) isometry that leaves EF unchanged.
   Also, $ABCD$ is congruent to each of the quadrilaterals shown in figures (a), (b), and (c). For each of these three quadrilaterals, propose a sequence of isometries that will transform $ABCD$ into it. Propose a few solutions. In each case try to find the solution that includes the use of as few steps as possible. Can you avoid using certain types of isometries? Which one(s)?

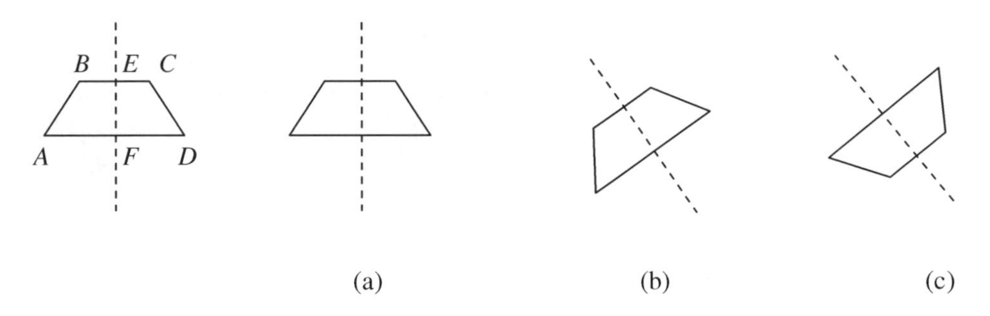

Solutions to Problem 2.

The solutions in each of the three cases are analogous to those of Problem 1 (a) and (b). We can always transform the given figure $ABCD$ in any congruent figure by means of isometries of type (i) and (ii). We shall never need a reflection in a plane since the figure does not change (is invariant) under a reflection in its *symmetry axis* (for $ABCD$ this axis is EF).

The diagram below illustrates a solution for the case (c). The transformation is performed as a translation followed by a rotation.

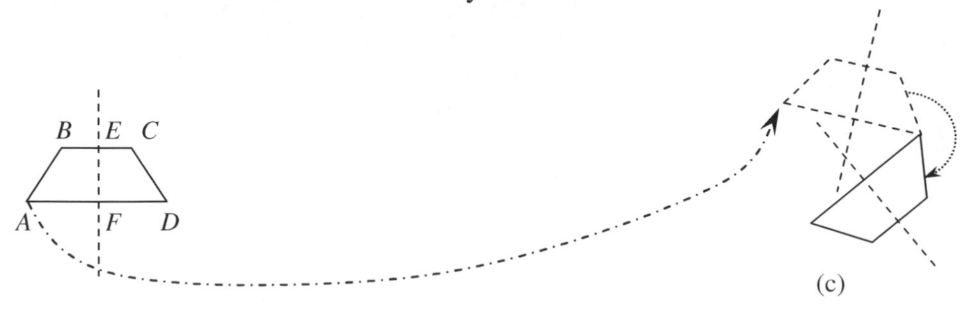

3. Prove that the radii of congruent circles are congruent.

Solution to Problem 3.

If two congruent circles are given, we can subject one of them to an isometry that will transform that circle into the second of these congruent circles, or, as we say in more common words, we shall make two circles to coincide.

If , as a result of such a transformation, their centers will coincide, then the distance from the centre to a point on the circle will be the same segment for both circles; hence in this case the radii will be congruent (they will just coincide).

Now let us show that the centers will necessarily coincide. We shall make it by means of a so-called *proof by contradiction*: we shall assume that the centers do not coincide and show that such an assumption leads to a contradiction with the facts that are known to be true (such as the axioms of Euclidean geometry).

Let us suppose that the centers of the circles did not coincide as a result of the isometry that makes the circles coincide: one centre is located at some point O and another at a different point P, as shown in the diagram below.

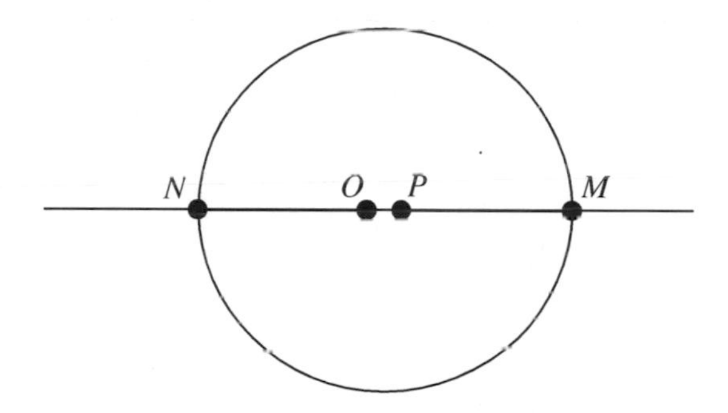

According to AXIOM 1, there is exactly one line that passes through the points O and P. Let M and N be the points of intersection of this line with the circle.

Since we have suggested that O is a centre of the circle, the segments NO and OM are congruent as two radii of the circle centred at O: $NO = OM$. Similarly, $NP = PM$ as two radii of the circle centred at P.

Point O lies in the interior of NP, hence $NO < NP$; similarly, $PM < OM$. Then we can write: $NO < NP = PM < OM = NO$, which means that $NO < NO$. This result contradicts to the axiom of congruence of segments which asserts that each segment is congruent to itself. Thus the suggestion that the centres do not coincide leads to a contradiction. Hence the centres must coincide, which means the congruence of the radii of the circles.

Test to Chapter 2.

1.

a) The proposition is given: *the absolute values of equal numbers are equal.* (i) Put it into conditional form. (ii) Formulate the converse, inverse, and contrapositive of this proposition. For each of them determine whether it is true or false.

b) It is known about the statements A and B that $\overline{B} \Rightarrow \overline{A}$ is true.
Is A sufficient for B? Is A necessary upon B? Substantiate your answers. Illustrate the situation by means of a Venn diagram

2. It is known about the statements A and B that A is sufficient but not necessary for B. For each of the following statements determine if possible whether it is true or false. (If it is impossible to determine, state: "undetermined" and explain why). Substantiate your answers; illustrate them by Venn diagrams.

(i) $\overline{B} \Rightarrow \overline{A}$

(ii) $B \Rightarrow A$

(iii) $B \Rightarrow \overline{A}$

(iv) B is necessary upon A.

(v) $\overline{A}$ is sufficient for $\overline{B}$

3. Prove the following corollaries of Common Notions (1-9):

(i) *multiples of equals are equal*:

$a = b, \quad \Rightarrow \quad na = nb$ for any natural number n

(ii) *equal parts of equals are equal*:

$$a = b, \quad \Rightarrow \quad \frac{a}{n} = \frac{b}{n}$$ for any natural number n.

Hints: for (i) use a *proof by induction* (see the discussion of The Principle of Mathematical Induction in the solution to problem 12 from section 1.2); for (ii) use the *proof by contradiction*.

Solutions to Chapter 2 Test.

1.

a) The proposition is given: *the absolute values of equal numbers are equal.* (i) Put it into conditional form. (ii) Formulate the converse, inverse, and contrapositive of this proposition. For each of them determine whether it is true or false.

SOLUTION.

(i) *If two numbers are equal, then their absolute values are equal.*
(Or to put it into more refined English: *if two numbers are equal, then so are their absolute values*).
One may also use a symbolic form:
$a = b \Rightarrow |a| = |b|$
The statement is true. The proof follows below.
Proof. It follows immediately from the definition of absolute value:

$$|a| = \begin{cases} a, & \text{if } a \geq 0 \\ -a, & \text{if } a < 0 \end{cases}.$$

If $a \geq 0$, then so is b, and $|a| = a = b = |b|$. Otherwise $a < 0$, then so is b, and $|a| = -a = -b = |b|$. In either case the absolute values are equal.
Let us notice that the absolute value of a number can be viewed as the distance between the point representing the number on the coordinate axis and the origin of the axis. Then, if $a = b$, these numbers are represented by the same point, so the distance from the origin is the same for a and b.

(ii) Converse: *If the absolute values of two numbers are equal, then the numbers are equal.* In symbolic form: $|a| = |b| \Rightarrow a = b$.
This statement is false, which can be proved by a *counterexample*:
$|-5| = 5 = |5|$; still $-5 \neq 5$.
(Notice that we cannot prove a statement *in general* by providing examples in support of it, but we can disprove a statement by providing a single *counterexample*.)

Inverse: *If two numbers are not equal, then their absolute values are not equal*, or: $a \neq b \Rightarrow |a| \neq |b|$.
It is false since it is logically equivalent to the converse statement. (Also, propose a counterexample)

Contrapositive: *If the absolute values of two numbers are not equal, then the numbers are not equal*, or: $|a| \neq |b| \Rightarrow a \neq b$.
It is true, which can be proved by contradiction: Let the absolute values of two numbers be unequal. If the numbers are equal, it implies the equality of their absolute values. Therefore, the equality of these numbers contradicts to the hypothesis; hence the numbers are not equal.
(We could just refer to the fact the contrapositive of a statement is logically equivalent to the statement.)

b) It is known about the statements A and B that $\overline{B} \Rightarrow \overline{A}$ is true. Is A sufficient for B? Is A necessary upon B? Substantiate your answers. Illustrate the situation by means of a Venn diagram.

SOLUTION.

$\overline{B} \Rightarrow \overline{A}$ means: if B is not true, then A is not true. It means that *B is necessary upon A*, i.e. $A \Rightarrow B$. The latter is the contrapositive of the given statement, and it can be read as *A is sufficient for B* (also: *A implies B*, or *B follows from A*).
The converse statement $B \Rightarrow A$ is not necessarily true, so *A is not necessary upon B*.

A diagram follows below. If a point is not located within the white disk ($\overline{B}$), it cannot be in the interior of the white one ($\overline{A}$), i.e. $\overline{B} \Rightarrow \overline{A}$.

Where would be located points illustrating the falsity of the inverse? converse? the validity of the contrapositive?

2.

It is known about the statements A and B that A is sufficient but not necessary for B. For each of the following statements determine if possible whether it is true or false. (If it is impossible to determine, state: "undetermined" and explain why). Substantiate your answers; illustrate them by Venn diagrams.

SOLUTION.

It is given: $A \Rightarrow B$ and $B \not\Rightarrow A$.
A corresponding diagram follows:

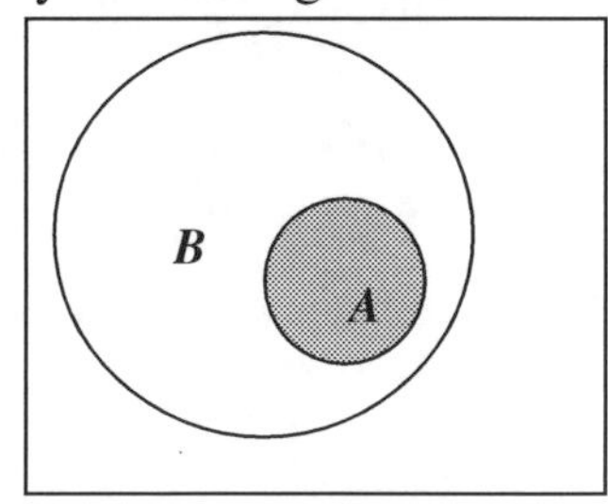

An illustration (or an illustration of a counterexample) for each statement is presented by the corresponding number located on the diagram below.

(i) $\overline{B} \Rightarrow \overline{A}$ TRUE (the contrapositive of $A \Rightarrow B$)

(ii) $B \Rightarrow A$ FALSE

(iii) $B \Rightarrow \overline{A}$ FALSE

(iv) B is necessary upon A. TRUE

(v) $\overline{A}$ is sufficient for $\overline{B}$. FALSE

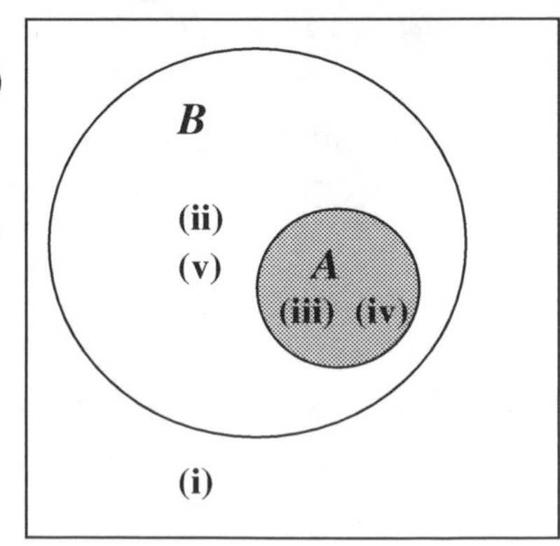

3. Prove the following corollaries of Common Notions (1-9):

(i) *multiples of equals are equal*:

$a = b, \Rightarrow na = nb$ for any natural number n

(ii) *equal parts of equals are equal*:

$$a = b, \Rightarrow \frac{a}{n} = \frac{b}{n} \text{ for any natural number } n.$$

Hints: For (i) use a *proof by induction* (see the discussion of The Principle of Mathematical Induction in the solution to problem 12 from section 1.2); for (ii) use the *proof by contradiction.*

SOLUTIONS.

(i)

First step: If $n = 1$ the statement is true since $a = b$ can be viewed as $1 \cdot a = 1 \cdot b$.
Inductive step: Suppose for some natural number n, it is true that $na = nb$. Since it is given that $a = b$, we can use Common Notion (1) to obtain: $na + a = nb + b$, which means that $(n+1)a = (n+1)b$. Thus we have proved, that if the statement holds for n, it holds for $(n+1)$ as well. This completes the inductive step; hence the statement is true for any natural number n.

(ii) Let us suppose that $a = b$, but $\frac{a}{n} \neq \frac{b}{n}$. We shall suggest for certainty that $\frac{a}{n} > \frac{b}{n}$ (the case $\frac{a}{n} < \frac{b}{n}$ can be considered in a similar manner).

Let $x = \frac{a}{n}$, and $y = \frac{b}{n}$; then, we supposed that $x > y$, which means that $x = y + d$, where d is some positive number.

According to the corollary (i) proved above, it follows from $x = y + d$ that $nx = n(y + d)$; hence $nx = ny + nd; \Rightarrow a = b + nd$, which cannot be true since by the hypothesis, $a = b$. Thus we have arrived to a contradiction that shows that our suggestion $\frac{a}{n} > \frac{b}{n}$ is false. Similarly, it can be shown that $\frac{a}{n} < \frac{b}{n}$ cannot be true.

Therefore, $\frac{a}{n} = \frac{b}{n}$, Q.E.D.

LIST 1. Basic results from CHAPTERS 1 – 4: Undefined Terms & First Axioms (neutral geometry); Definitions, Angles; Congruence of Triangles.

Undefined notions: point, line, plane, congruence of segments.

AXIOM 1. ***Given any two points in space, there is one and only one line passing through these points.***

AXIOM 2

(i) ***Each segment is congruent to itself:*** $a = a$ ***(reflexivity);***

(ii) ***If one segment is congruent to the other, then the second segment is congruent to the first one:*** $a = b \Rightarrow b = a$ ***(symmetry);***

(iii) ***If each of two segments is congruent to the third segment, they are also congruent to each other:*** $\left.\begin{array}{l} a = c \\ b = c \end{array}\right\} \Rightarrow a = b$ ***(transitivity).***

AXIOM 3 ***One and only one segment congruent to a given segment can be laid off on a ray starting from the origin.***

Definition *The segment connecting given two points is called the distance between these points.*

Definition *An isometry is a distance–preserving transformation of figures.*

Definition *Two geometric figures are said to be congruent if one of them can be made to coincide with the other by means of an isometry.*

AXIOM 4 (Existence of Isometries).

(i) For any pair of points, there exists an isometry that carries one of these points into the other.

(ii) For any pair of rays with a common vertex, there exists an isometry that carries one of these rays into the other.

(iii) For any line in a plane, there exists an isometry that leaves every point of the line unchanged and maps a half-plane lying on one side of the line onto the half-plane on the opposite side of the line.

Definition *A locus is a set of points such that*

(i) *all points in the set satisfy a given condition, and*

(ii) *all points that satisfy the given condition belong to this set.*

Definition *A circle is the* **locus** *of points equidistant from a given point called the centre of the circle.*

Definition *A line that touches a circle in but one point is called a tangent; this common point is called a point of tangency* (sometimes – *a point of contact*).

Theorem 2.2.1 **Congruent arcs (of the same circle) are subtended by congruent chords** can be formulated as ***if two arcs of a circle are congruent, then the chords that subtend them are congruent as well.***

Theorem 2.2.2. ***A circle cannot have more than one centre.***

Definition An *angle* is a figure formed by two rays that emanate from a common point and include a part of the plane. The point is called the *vertex* of the angle, the rays are called the *sides* or *arms* of the angle, and the included part of the plane is called the *interior* of the angle, whereas the rest of the plane is called the *exterior*.

Definition Two angles are called *adjacent* (in some literature *adjoining*) if they share a common vertex and one arm, and their interiors are located on the opposite sides of the common arm. If non-common arms of two adjacent angles form a straight line, the angles are said to be *supplementary*.

Definition A *right angle* is an *angle that is equal to its supplement.*

Definition Two straight lines (rays, segments) are *perpendicular* to each other if they form a right angle.

Theorem 3.2.1. ***Right angles exist.***

Theorem 3.2.2 ***All right angles are congruent.***

Theorem 3.2.3 ***One and only one perpendicular can be erected from a given point in a given line.***

Theorem 3.2.4. ***A perpendicular can be dropped onto a given line from a given exterior point.***

Theorem 3.3.1. ***Any two vertical angles are congruent.***

Theorem 3.4.1 ***In the same circle or in congruent circles:***

(i) ***if two central angles are congruent, then the corresponding arcs are congruent;***

(ii) ***if two arcs are congruent, then the corresponding central angles are congruent.***

Theorem 4.2.1. ***An equilateral triangle with a side congruent to a given segment can be constructed.***

Theorem 4.2.2. ***A Given segments l and b such that b<2l, an isosceles triangle with b as a base and l as a lateral side can be constructed.***

Definition Two points are said to be *symmetric in a line* if they are located in the opposite half-planes formed by the line, on a perpendicular to the line and are equidistant from the foot of the perpendicular.

Definition *Corresponding parts* are two parts of congruent figures that would coincide if the figures were made to coincide by an isometry.

Theorem 4.4.1 ***(SAS condition). If two triangles have two sides and the included angle of the one equal, respectively, to two sides and the included angle of the other, the triangles are congruent.***

Corollary 1. ***Two right triangles are congruent if two legs of one are equal respectively to two legs of the other.***

Theorem 4.4.2 ***(ASA condition) If two triangles have two angles and the included side in the one congruent, respectively, to two angles and the included side in the other, the triangles are congruent.***

Corollary 1. ***Two right triangles are congruent if a leg and the angle formed by this leg and the hypotenuse in one triangle are congruent, respectively, to a leg and the angle formed by this leg and the hypotenuse in the other triangle.***

Theorem 4.4.3 ***In an isosceles triangle, the angles opposite the congruent sides are congruent.***

Theorem 4.4.4 ***If two angles in a triangle are congruent, the sides opposite to these angles are congruent, i.e. the triangle is isosceles.***

Theorem 4.4.5 ***(SSS condition). If two triangles have three sides of the one congruent respectively to the three sides of the other, the triangles are congruent.***

Theorem 4.4.6. ***Every angle has a bisector.***

Theorem 4.4.7 ***An angle has exactly one bisector.***

Theorem 4.4.8 ***In an isosceles triangle the bisector of the angle included by the congruent sides is also a median and an altitude.***

Corollary 2. ***The bisector of the angle at the vertex of an isosceles triangle is a symmetry axis of the triangle.***

Theorem 4.5.1. ***An exterior angle of a triangle is greater than either of the opposite interior angles.***

Corollary 1. ***If in a triangle one of its angles is right or obtuse, then the other two angles are acute.***

Corollary 2. ***Given a line and a point, there exists only one perpendicular to the given line through the given point.***

Corollary 3, ***Given a line, there exists exactly one isometry that reflects the plane in this line, i.e. for a given line, type (iii) isometry is unique.***

Corollary 4. ***For every point there exists exactly one point symmetric to it in a given line.***

Theorem 4.5.2 ***(AAS condition) If two triangles have two angles and the side opposite to one of these angles in the one congruent, respectively, to two angles and the corresponding side in the other, the triangles are congruent.***

Theorem 4.6.1. ***In any triangle***

(i) ***if two sides are congruent, the opposite angles will be congruent;***

(ii) ***if two sides are not congruent, the angle opposite to the greater side will be greater than the angle opposite to the smaller side.***

Theorem 4.6.2. ***In any triangle***

(i) ***if two angles are congruent, the opposite sides are congruent;***

(ii) ***if two angles are not congruent, the side opposite to the greater angle will be greater than the other.***

Corollary 1. ***An equiangular triangle is also equilateral.***

Corollary 2. ***In a right triangle the hypotenuse is the greatest side.***

Theorem 4.7.1 ***(Triangle Inequality). The sum of any two sides of a triangle is greater than the third side.***

Corollary 1. ***Each side of a triangle is greater than the difference between the other two sides.***

Theorem 4.7.2 ***A straight segment connecting two points is less than any broken line connecting these points.***

Theorem 4.7.3 ***If two triangles have two sides of the one equal to two sides of the other, then ,***

(i) if the included angles are unequal, then the third sides are unequal, the greater side being opposite the greater angle; and conversely,

(ii) if the third sides are unequal, then the angles opposite those sides are unequal, the greater angle being opposite the greater side.

Theorem 4.8.1. ***A perpendicular drawn from an external point to a given line is less than any oblique drawn to the same line from the same point.***

Theorem 4.8.2. ***If a perpendicular and two obliques are drawn to a line from the same exterior point, then***

(i) if the feet of the obliques are equidistant from the foot of the perpendicular, then the obliques are congruent;

(ii) if the feet of the obliques are not equidistant from the foot of the perpendicular, the oblique whose foot is farther from the foot of the perpendicular, is greater than the other oblique.

Theorem 4.8.3. ***If a perpendicular and two obliques are drawn from an exterior point to a line, then***

(i) if the obliques are congruent, then their feet are equidistant from the foot of the perpendicular

(ii) if the obliques are not congruent, the foot of the greater one will be farther from the foot of the perpendicular.

Corollaries from ***SAS and ASA conditions:***

Two right triangles are congruent if

(i) two legs of the one are equal, respectively, to the two legs in the other;

(ii) an acute angle and its adjacent leg in the one are equal, respectively, to an acute angle and its adjacent leg in the other;

Theorem 4.9.1 ***Two right triangles are congruent if the hypotenuse and a leg of the one are, respectively, equal to the hypotenuse and a leg of the other.***

Theorem 4.9.2 ***Two right triangles are congruent if an acute angle and the hypotenuse of the one are congruent respectively to an acute angle and the hypotenuse of the other.***

Theorem 4.10.1a. ***A point located on the perpendicular bisecting a segment is equidistant from the endpoints of the segment.***

Theorem 4.10.2a. ***If some point is equidistant from the endpoints of a segment, then this point is located on the perpendicular bisecting the segment.***

Theorem 4.10.3a. ***If a point is not located on the perpendicular bisecting a segment, it is not equidistant from the endpoints of the segment.***

Corollary (a). ***The perpendicular bisecting a segment is a locus of points equidistant from the endpoints of the segment.***

Theorem 4.10.1b. ***A point located on the bisector of an angle is equidistant from the sides of the angle.***

Theorem 4.10.2b. ***If some point is equidistant from the sides of the angle, then this point is located on the bisector of the angle.***

Theorem 4.10.3b. ***If a point is not located on the bisector of an angle, it is not equidistant from the sides of the angle.***

Corollary (b). ***The bisector of an angle is a locus of points equidistant from the sides of the angle.***

LIST 2. Basic results from CHAPTERS 5 – 7: Constructions in Neutral Geometry; Parallel Lines; Parallel Postulate; Parallelogram and Trapezoid.

Constructions:

a) To construct an angle congruent to a given angle, with the vertex at a given point and one side lying in a given line.
b) To construct a triangle having given the three sides (*SSS*).
c) To construct a triangle having given two sides and the included angle (*SAS*).
d) To construct a triangle having given two angles and the included side (*ASA*).
e) To construct a triangle having given two sides and an angle opposite to one of them (*SSA*).
f) To bisect an angle (draw the symmetry axis of an angle).
g) To erect a perpendicular to a given line from a given point in the line.
h) To drop a perpendicular onto a line from a given exterior point.

i) To draw a perpendicular bisecting a given segment (the symmetry axis of a segment).

Theorem 6.1.1. ***Two perpendiculars to the same line are parallel.***

Theorem 6.2.1. ***If two lines are cut by a common transversal so as to make***

(i) two corresponding angles equal; or
(iii) two alternate angles equal; or

(iv) two interior (or two exterior) angles lying on the same side of the transversal supplementary,

then in each case the two lines are parallel.

Problem 1. *Through a given (external) point draw a line parallel to a given line.*

AXIOM 5 (Parallel Postulate) ***Only one line can be drawn through a given point parallel to a given line.***

Corollary 1. ***If a line is transversal to one of two parallel lines it will be also transversal to the other.***

Corollary 2. ***Lines parallel to the same line are parallel to each other.***

Theorem 6.2.2. ***If two parallel lines are cut by a common transversal, then***

(i) each exterior (interior) angle is equal to its corresponding;

(ii) the alternate angles are equal to one another;

(iii) the interior (exterior) angles on the same side of the transversal are supplementary.

Corollary 1. **Given two parallel lines, a perpendicular to one of them is also a perpendicular to the other.**

Theorem 6.3.1. ***If two lines are cut by a common transversal, and***

(i) the corresponding angles are not equal, or

(ii) the alternate angles are not equal, or

(iii) two interior (or exterior) angles lying on the same side of the transversal are not supplementary,

then in each case the two lines are not parallel.

Theorem 6.3.2. ***If two nonparallel lines are cut by a common transversal, then***

- ***(i)*** ***the corresponding angles are not equal;***
- ***(ii)*** ***the alternate angles are not equal;***
- ***(iii)*** ***the interior [exterior] angles on the same side of the transversal are not supplementary.***

Parallel postulate (Euclidean formulation):

If a straight line falling on two straight lines makes the interior angles on the same side less than two right angles, the straight lines, if produced indefinitely, meet on that side on which are the angles less than the two right angles.

Theorem 6.3.3. ***A perpendicular and an oblique drawn to the same line, are not parallel.***

Theorem 6.3.4. ***If two lines intersect, their respective perpendiculars are not parallel.***

Theorem 6.4.1. ***Two angles whose sides are respectively parallel are either equal or supplementary.***

Theorem 6.4.2. ***Two angles whose sides are respectively perpendicular to each other are either equal or supplementary.***

Theorem 6.5.1. ***The sum of the three angles of triangle is equal to two right angles.***

Corollary 1. ***An exterior angle of a triangle is equal to the sum of its opposite (non-adjacent) interior angles.***

Corollary 2. ***If two angles of a triangle are equal respectively to two angles of another triangle, then the third angles are also respectively equal.***

Corollary 3. ***In any right triangle the two acute angles are complementary.***

Corollary 4. ***In a right isosceles triangle each of the acute angles measures 45°.***

Corollary 5. ***Each angle of an equilateral triangle measures 60°.***

Corollary 6. ***In a right triangle with one of its acute angles equal 30°, the leg opposite to this angle equals half of the hypotenuse.***

Corollary 7. ***The sum of the angles of any quadrilateral is equal to four right angles.***

Theorem 6.5.2. ***The sum of the interior angles of a convex polygon with n sides is equal to two right angles repeated n-2 times.***

Theorem 6.5.3. ***The sum of the interior angles of a polygon with n sides is equal to two right angles repeated n – 2 times.***

Theorem 6.5.4. ***For any convex polygon the sum of the exterior angles (obtained by producing each side beyond one vertex, in order) is equal to four right angles.***

Theorem 6.6.1. ***Given a centre of symmetry, each point in the plane has exactly one symmetric image about the centre.***

Theorem 6.6.2. ***If for two points (A and B) their symmetric images (A´ and B´) about a given centre (O) are constructed, then***

(i) ***the segment joining the points (A and B) will be parallel and equal to the segment joining their images (A´ and B´);***

(ii) for each point of the segment connecting the given points (A and B), its symmetric image lies on the segment (A′B′) connecting the images of the given points.

Theorem 6.6.3. ***Symmetry in a point is an isometry.***

Corollary 1 ***If two figures are symmetric in a point, they are congruent.***

Theorem 7.1.1 ***In a parallelogram opposite sides are congruent, opposite angles are congruent, and two angles with a common side are supplementary.***

Corollary 1. ***If two lines are parallel, the distance from any point of one line to the other line will be the same for all points of the first line.***

Theorem 7.1.2. ***If a convex quadrilateral has***

(i) both pairs of opposite sides congruent, or

(ii) one pair of sides congruent and parallel,

then the quadrilateral is a parallelogram.

Theorem 7.1.3. ***The diagonals of a parallelogram bisect each other.***

Theorem 7.1.4. ***If in a quadrilateral diagonals bisect each other, the quadrilateral is a parallelogram.***

Theorem 7.1.5. ***In a parallelogram the point of intersection of the diagonals is the centre of symmetry of the parallelogram.***

Theorem 7.1.4. ***In a rectangle***

(i) diagonals are congruent;

(ii) each of two lines passing through the centre of symmetry of the rectangle parallel to its sides, is an axis of symmetry of the rectangle.

Theorem 7.2.2 ***In a rhombus,***

i) the diagonals are mutually perpendicular and bisect the angles of the rhombus;

ii) each diagonal is an axis of symmetry of the rhombus.

Theorem 7.3.1. ***If three or more parallel lines cut off equal segments on one transversal, they cut off equal segments on any other transversal.***

Corollary 1. ***The line drawn through the midpoint of one side of a triangle parallel to another side, bisects the third side of the triangle.***

Theorem 7.3.2. ***(Properties of a midline of a triangle). The segment joining the midpoints of two sides of a triangle is parallel to the third side and congruent to half the third side.***

Theorem 7.4.1. ***The midline of a trapezoid is parallel to its bases and equal half of the sum of the bases.***

Construction problem. *Divide a given segment into a given number of equal segments.*

Transformation of a figure is called a *translation* if it shifts each point of the figure through the same segment in the same direction parallel to a fixed straight line.

Theorem 7.5.1. ***A translation is an isometry.***

Corollary 1. ***A translation transforms a figure into a congruent figure.***

Two successive transformations (of the plane) result in a transformation called a *composition* of the two.

Theorem 7.5.2. ***A composition of (two) translations is a translation.***

LIST 3: Basic statements and problems from CHAPTERS 8, 9, 10, and 11: Circles; Similarity; Regular Polygons and Circumference; Areas.

Theorem 8.1.1. *There is one and only one circle passing through three given noncollinear points.*

Corollary 1. *Three perpendicular bisectors drawn to the three sides of a triangle, are concurrent.*

Theorem 8.1.2. *A diameter that is perpendicular to a chord, bisects the chord and its subtended arcs.*

Theorem 8.1.3 *(i) A diameter that bisects a chord is perpendicular to the chord and bisects the arc subtented by the chord.*

(ii) A diameter that bisects an arc is a perpendicular bisector of the chord that subtends the arc.

Theorem 8.1.4. *Two arcs included between a pair of two parallel chords, are congruent.*

Problem 1. *Bisect a given arc of a circle.*

Problem 2. *Find the centre of a circle.*

Theorem 8.2.1. *In the same circle, or in congruent circles,*

(i) *equal arcs are subtended by equal chords that are equidistant from the centre;*

(ii) *of two unequal arcs, each less than semicircle, the greater is subtended by a greater chord that is nearer to the centre than the shorter chord.*

Theorem 8.2.2. *In the same circles, or in congruent circles,*

(i) *equal chords are equidistant from the centre and subtend equal arcs;*

(ii) *chords that are equidistant from the centre are equal and subtend equal arcs;*

(iii) *of two unequal chords, the longer is nearer the centre and subtends the greater arc;*

(iv) *of two chords that are unequal distances from the centre, the one nearer the centre is greater and subtends the greater arc.*

Theorem 8.2.3. *A diameter is the greatest chord in a circle.*

Theorem 8.3.1. *If a line is perpendicular to a radius of a circle at the endpoint of the radius lying on the circle, then the line is tangent to the circle,* and conversely:

Theorem 8.3.2. *If a line is tangent to a circle, the radius drawn to the point of tangency is perpendicular to the line.*

Theorem 8.3.3. *If a tangent is parallel to a chord, then the point of tangency bisects the arc subtended by the chord.*

Theorem 8.4.1. *If two circles have a common point that is exterior to their centre line, then they have one more common point, symmetric in their centre line with the first common point.*

Theorem 8.4.2. *If a common point of two distinct circles lies in their centre line, the circles are tangent.*

Theorem 8.4.3. (converse to 8.4.2) *If two circles touch each other, the point of tangency lies in their centre line.*

Corollary 1. *If two circles touch each other, they have a common tangent at the point of contact.*

Theorem 8.5.1. *An inscribed angle is half the central angle subtended by the same arc.*

Corollary 1. (Another formulation of the theorem.) *An inscribed angle is measured by one half of its intercepted arc.*

Corollary 2. *All angles inscribed in the same arc or in congruent arcs are congruent.*

Corollary 3. *An angle inscribed in a semicircle (an angle standing on a diameter) is a right angle.*

Construction problems.

1. Construct a right triangle, given its hypotenuse, *c*, and one leg, *b*.
2. *Erect a perpendicular to a ray at its vertex, without extending the ray.*
3. *Draw a tangent through a given point to a given circle.*

Corollary. *Two tangents drawn from an exterior point to a circle form equal angles with the line connecting this point with the centre of the circle.*

Problem 4. *Given two circles, draw their common tangents.*

Theorem 8.5.2. *An angle with the vertex lying inside a circle is measured by one half of the sum of the arcs intercepted by its arms and extensions of its arms.*

Lemma 8.5.3. *An angle formed by a chord and a tangent is measured by one half the arc intercepted by the chord.*

Theorem 8.5.4. *An angle formed by two secants, two tangents, or a secant and a tangent, that intersect outside the circle, is measured by one half the difference of the intercepted arcs.*

Problem 5. *Given a straight segment as a chord, construct a segment (of a circle) in which a given angle is inscribed.*

Theorem 8.7.1 *One and only one circle can be circumscribed about a triangle.*

Theorem 8.7.2 *One and only one circle can be inscribed in a triangle.*

Corollary. *Three bisectors of a triangle meet at one point (are concurrent).*

Theorem 8.7.3. *A convex quadrilateral can be inscribed in a circle iff (if and only if) its opposite angles are supplementary*

Corollary 1. *A parallelogram is inscriptible only if it is a rectangle*

Corollary 2. *A trapezoid is inscriptible only if it is isosceles.*

Theorem 8.7.4. *In an inscriptible quadrilateral the sums of the opposite sides are equal to each other.*

Theorem 8.8.1. *The altitudes of a triangle meet in a point.*

Theorem 8.8.2. *The medians of a triangle meet in a point which is two thirds of the distance from any vertex to the middle point of the opposite side.*

Theorem 9.1.1 *If the least of two segments is contained an integer number of times without a remainder in the greatest segment, then the least segment is the greatest common measure of the two segments.*

Theorem 9.1.2 *If the least of two segments is contained an integer number of times, with a remainder, in the greatest segment, then their greatest common measure, if it exists, is also the greatest common measure of the least segment and the remainder.*

I CONTINUITY AXIOM (Archimedes' Axiom). *For any two segments:*

either one of them is a multiple of the other,

or there exists such a number n that the greatest segment is less than n-multiple of the least segment.

Theorem 9.1.3. *A diagonal of a square is incommensurable with its side.*

Theorem 9.1.4 *To every segment there correspond a positive real number that represents its length.*

II CONTINUITY AXIOM (Cantor's Principle of Nested Segments). *For any sequence of nested segments there exists a point common to all of them.*

Corollary. *If in a sequence of nested segments there is a segment lesser than any given segment, then there is one and only one point common to the sequence of nested segments.*

Theorem 9.1.5 *Every real positive number is the length of some segment.*

Definition. *Two triangles are said to be similar if.*

(i) *the angles of one triangle are respectively equal to the angles of the other;*

(ii) *the sides of one are proportional to the corresponding sides of the other.*

The following lemma proves the existence of similar triangles.

Lemma 9.2.1. *A line parallel to a side of a triangle, cuts off a triangle similar to the given one.*

Theorem 9.3.1. *Two triangles are similar if*

(i) two angles of one triangle are respectively equal to two angles of the other triangle;

or

(ii) an angle of one is equal to an angle of the other and the including sides are proportional; or

(iii) all their sides are respectively proportional

Theorem 9.4.1. *Two right triangles are similar if*

(i) *an acute angle of one is equal to an acute angle of the other, or*

(ii) *the legs of one are proportional to the legs of the other, or*

(iii) *a leg and the hypotenuse of one are proportional to a leg and the hypotenuse of the other.*

Theorem 9.4.2. *The altitudes of two similar triangles are proportional to the bases.*

Theorem 9.5.1. *Each of similar polygons can be decomposed into the same number of similar and similarly situated triangles.*

Theorem 9.5.2. *The perimeters of similar polygons are proportional to any of their corresponding sides.*

Prop. 9.6.1. *An isometric image of a figure is a figure similar to the original one,* or briefly: *congruent figures are similar (with the similarity ratio $k = 1$).*

Prop. 9.6.2. *A composition of an isometry and a homothety is a similarity transformation.*

Prop. 9.6.3. *A composition of two homotheties is a similarity transformation.*

Theorem 9.6.4. (i) A figure similar to a circle is a circle and

(ii) any two circles are similar,

i.e. *all circles and only them are similar to each other.*

Problem 9.7.1. Construct the triangle given its angle C, the ratio of the sides including this angle, and the altitude, h, dropped from vertex C onto the opposite side.
Problem 9.7.2. Inscribe a circle in a given angle so as to make the circle pass through a given point.
Problem 9.7.3. Given a triangle, ΔABC, inscribe in this triangle a rhombus with a given acute angle and one of its (rhombus) sides lying on the base, AB, whereas two vertices of the rhombus lie on sides AC and BC.
Theorem 9.8.1. If a series of parallel lines intersects two sides of an angle, the corresponding segments cut on the two sides by the parallel lines are proportional.
Theorem 9.8.2. If two parallel lines are met by a pencil (bundle) of lines, the corresponding segments cut off on the parallel lines are proportional.
Problem 9.8.1. *Given a segment, divide it into three parts in a given ratio $m:n:p$ where m,n,p are given numbers or segments.*

Problem 9.8.2. *Find the fourth proportional to three given segments,* i.e. having given three segments, a, b, c, find such a segment x that $\frac{a}{b} = \frac{c}{x}$.

Theorem 9.8.3. The bisector of an angle of a triangle divides the opposite side into two segments which are proportional to the sides including the angle.

Theorem 9.8.4. (the property of an exterior angle of a triangle). The bisector of an exterior angle of a triangle meets the extension of the opposite side at such a point that the distances from that point to the endpoints of the opposite side are proportional to the two respective sides of the triangle ending at these points.
Lemma 9.9.1. *In a right triangle, the altitude drawn to the hypotenuse divides the triangle into two similar triangles, and each of them is similar to the original triangle.*

Theorem 9.9.2. In a right triangle,

(i) the altitude drawn to the hypotenuse is the mean proportional between the segments into which it divides the hypotenuse, and

(ii) each leg is the mean proportional between its projection onto the hypotenuse and the hypotenuse.

Corollary 1 In a circle,

(i) a perpendicular dropped from a point on a circle onto a diameter is the mean proportional between the segments into which it divides the diameter, and

(ii) a chord connecting a point on a circle with an endpoint of a diameter is the mean proportional between the diameter and the projection of the chord onto the diameter.

Theorem 9.9.3. (Pythagorean Theorem) *If the hypotenuse and legs of a right triangle are measured by a common measure, the square of the length of the hypotenuse is equal to the sum of the squares of lengths of the legs.*

Theorem 9.9.4. *In a right triangle, the ratio of squares of its legs is equal to the ratio of the corresponding projections of the legs upon the hypotenuse.*

Remark 1. We have established that in a right triangle its legs, a, b, hypotenuse, c, the altitude, h, drawn to the hypotenuse, and the projections, a', b', of the legs onto the hypotenuse are connected by the following relations:

1. $a'c = a^2$,
2. $b'c = b^2$,
3. $a^2 + b^2 = c^2$,
4. $a' + b' = c$,
5. $h^2 = a'b'$.

Theorem 9.9.6. *In any obtuse triangle the square of the side opposite the obtuse angle is equal to the sum of the squares on the other two sides increased by twice the product of one of those sides and the projection of the other upon it.*

Corollary 1 *The square of a side of a triangle is less, equal, or greater than the sum of the squares of the two other sides if the angle opposite to that side is, respectively, acute, right, or obtuse.*

Corollary 2 *An angle of a triangle is acute, right, or obtuse, depending on whether the square of the opposite side is less, equal, or greater than the sum of the other two sides.*

Theorem 9.9.7. *The sum of the squares of the diagonals of a parallelogram is equal to the sum of the squares of the four sides.*

Problem. *Given the sides of a triangle, find its altitudes.*

Theorem 9.10.1. *If in a circle a chord and a diameter intersect each other, the point of intersection divides them into segments such that the product of the segments of the chord equals to the product of the segments of the diameter.*

Corollary 1. *Given a point in the interior of a circle, the product of segments into which the point divides chords passing through the point, is invariant (i.e., it is the same for all chords passing through that point).*

Theorem 9.10.2. *Given a circle and a point in the exterior of the circle, for any secant emanating from that point the product of the segments connecting the given point with the points where the secant meets the circle is equal to the square of the segment of the tangent drawn to the circle from the same point.*

Corollary 1. *For any secant drawn to a circle from an exterior point, the product of the secant and its exterior part is invariant. (This invariant value equals to the square of the tangent drawn to the circle from the same point.)*

Theorem 9.11.1 (*Law of cosines*) *In a triangle with the sides* a, b, *and* c, *the square of the side opposite* $\angle C = \gamma$ *is expressed as*

$c^2 = a^2 + b^2 - 2ba\cos\gamma$ *if* $\angle C$ is acute, and

$c^2 = a^2 + b^2 + 2ba\cos(180° - \gamma)$ *if* $\angle C$ *is obtuse.*

Theorem 9.11.2. (*Law of sines*). *In any triangle with the sides* a, b, c, *opposed the angles* α, β, γ, *respectively, and the circumradius R,* $\dfrac{a}{\sin\alpha} = \dfrac{b}{\sin\beta} = \dfrac{c}{\sin\gamma} = 2R$.

Problem 9.12.1. *Given a segment, divide it into extreme and mean ratio, i.e. divide the segment into two parts such that the ratio of the greater part to the lesser part is equal to the ratio of the whole segment to the greater part.*

Theorem 10.1.1. If a circle is divided into (more than two) equal parts, then

(i) the figure obtained by joining the neighbouring points of the partition by consecutive chords, is a regular polygon (inscribed in the circle);

(ii) the figure obtained by drawing tangents touching the circle at the points of the partition and extending each of these tangents till its intersection with the neighbouring two, is a regular polygon (circumscribed about the circle).

Theorem 10.1.2. If a polygon is regular, then

(i) a circle can be circumscribed about it;

(ii) a circle can be inscribed in it.

Corollary 1. *The incentre and circumcentre of a regular polygon coincide.*

Corollary 2. *To find the incentre/circumcentre of a regular polygon, one can*

(i) find the point of intersection of the perpendiculars bisecting any two sides of the polygon, or

(ii) find the point of intersection of the bisectors of two (interior) angles of the polygon, or

(iii) find the point of intersection of a bisector of an (interior) angle of the polygon with the perpendicular bisecting a side of the polygon.

Corollary 3. *A bisector of an interior angle of a regular polygon is a symmetry axis of the polygon.*

Corollary 4. *A perpendicular that bisects any side of a regular polygon is a symmetry axis of the polygon.*

Corollary 5. *For a regular polygon with even number of sides, its incentre (circumcentre) is the centre of symmetry.*

Theorem 10.1.3. Any two regular polygons with the same number of sides are similar, and their sides, their apothems, and their radii are in the same ratio.

Corollary 1. *The perimeters of two regular polygons are in the same ratio as their radii or apothems.*

Problem 10.1.1. *Given a circle of radius* R, *find the side of an inscribed*

(i) square,

(ii) regular hexagon,

(iii) regular triangle.

Problem 10.1.2. *Double the number of sides of an inscribed regular polygon, i.e.*

(i) Given a regular polygon with n *sides inscribed in a circle, inscribe in the circle a regular polygon with* $2n$ *sides;*

(ii) Express a_{2n} *in terms of* a_n *and* R.

Problem 10.1.3. *Inscribe a regular decagon in a given circle and express its side in terms of the radius.*

Lemma 10.2.1. The sequence of the perimeters of inscribed convex regular polygons obtained by doubling the number of sides of a polygon is monotonically increasing.
Theorem 10.2.2. The sequence of the perimeters of inscribed convex polygons is increasing if each subsequent polygon is obtained from the preceding one by adding new vertices.

Lemma 10.2.3. The sequence of the perimeters of circumscribed convex regular polygons obtained by doubling the number of sides of a polygon is monotonically decreasing.
Corollary 1. The perimeter of any regular polygon circumscribed about a circle of radius R, is less than $8R$ if the polygon has more than four sides and is obtained as a result of sequential doubling the number of sides of a regular triangle or a square.
Theorem 10.2.4. The sequence of the perimeters of circumscribed convex polygons is decreasing if each subsequent polygon is obtained from the preceding one by adding new sides that are tangent to the circle.

Lemma 10.2.5. A convex broken line is less than any broken line that includes it.
Lemma 10.2.6. The perimeter of any convex polygon is less than the perimeter of a polygon that includes the first polygon.
Corollary 1. The perimeter of any convex polygon inscribed in a circle is less than the perimeter of a polygon circumscribed about this circle.
Corollary 2. The perimeter of any convex polygon inscribed in a circle of radius R, is less than $8R$.
Corollary 3. The perimeter of any convex regular polygon circumscribed about a circle of radius R, is less than $16R$.

Lemma 10.2.7. A side of a circumscribed regular polygon can be made less than any given segment by choosing a polygon with a sufficiently large number of sides.
Lemma 10.2.8. For a given circle, the difference between the perimeters of regular polygons with the same number of sides, one of them inscribed and the other circumscribed, can be made less than any given segment by sequential doubling the number of sides of the polygons.
Theorem 10.2.9. For any circle, there exists a unique segment that is greater than the perimeter of any inscribed regular polygon and less than the perimeter of any circumscribed regular polygon, where the polygons are obtained by sequential doubling their number of sides. This segment is called the *circumference of the circle.*
Theorem 10.2.10. The circumference of a circle is proportional to its diameter, and the coefficient of proportionality is universal (the same) for all circles.

Theorem 10.3.1. If a sequence has a limit, this limit is unique.
Theorem 10.3.2. (Weierstrass' Theorem). If a sequence is monotonically increasing and bounded above, it has a limit.
Theorem 10.3.3. The sequence of the lengths of the perimeters of regular inscribed polygons obtained by doubling the number of their sides converges to a certain limit.
By definition, this limit, C, is the lengths of the circumference of the considered circle.

Area: Basic suggestions (principles).

By the *area* of a figure we mean the amount of surface contained within its bounding lines. Our problem is to find numerical expressions for the areas of various figures. We require that numerical expressions of areas satisfy the following principles (axioms):

(1) *The numbers that represent areas of congruent figures are equal.*
(2) *If a figure is decomposed into a few closed figures, then the number representing the area of the whole figure is equal to the sum of the numbers representing the areas of all parts of the decomposition.*

Lemma 11.1.1. If a side of one square is n times (n is an integer) the side of some other square, the area of the first square is n^2 times the area of the second square.
Theorem 11.1.2. The area of a rectangle equals to the product of its measurements.

Theorem 11.1.3. *The area of a parallelogram equals to the product of its height by its base:*

$\mathcal{A} = h \cdot b.$

Theorem 11.1.4. *The area of a triangle equals half of the product of its height by the base:*

$\mathcal{A} = ½\, hb.$

Corollary 1. *If a side and the corresponding altitude of one triangle are, respectively, congruent to a side and the corresponding altitude of the other, then the triangles are equivalent* (even though they are not congruent!).

Corollary 2. *The area of a right triangle is half the product its legs.*

Corollary 3. *If an angle of a triangle is congruent to an angle of some other triangle, the areas of the two triangles are in the same ratio as the products of the sides including congruent angles.*

Corollary 4. *The area of a rhombus is half the product its diagonals.*

Theorem 11.1.5. *The area of a trapezoid equals the product of its height by half-sum of its bases.*

Corollary 1. *The area of a trapezoid equals the product of its midline by its height.*

Theorem 11.1.6. *If a polygon is circumscribed about a circle, the area of the polygon equals half the product of its perimeter and the radius of the circle:* $\mathcal{A} = ½\, P\, r$.

Corollary 1. *The area of a regular polygon is half the product its perimeter and the apothem.*

Problem 11.1.1. (Heron's Formula). If the sides of a triangle are a, b, c, its area equals $\mathcal{A} = \sqrt{p(p-a)(p-b)(p-c)}$, where $p = ½P = ½(a+b+c)$.

Theorem 11.2.1. *The areas of two similar triangles are to each other in the same ratio as the squares of (any two) corresponding sides.*

Theorem 11.2.2. *The areas of two similar polygons are to each other in the same ratio as the squares of any two corresponding sides or diagonals.*

Corollary 1. *The areas of two regular polygons with the same number of sides are in the same ratio as the squares of their sides, or their apothems, or circumradii.*

The results of this section (11.2) can be briefly reformulated as follows:

The ratio of the areas of two similar figures equals k^2, where k is their similarity ratio.

Lemma 11.3.1. *When the number of the sides of a regular polygon doubles indefinitely, its side tends to zero and its apothem tends to the radius.*

$$\mathcal{A}\ (\text{circle}) = \pi R^2.$$

Corollary 1. *The areas of two circles are in the same ratio as the squares of their radii (diameters).*

Corollary 2. *The area of a sector is half the product of its arc and the radius.*

If the angle measures θ radians, the length of the arc is θR, and the area of the sector is

$$\mathcal{A}\ (\text{sector}) = \theta R \cdot \frac{R}{2} = ½\, \theta R^2.$$